T0275576

CAMBRIDGE LECTURE NOTES IN PHYSICS 10
General Editors: P. Goddard, J. Yeomans

Self-Organized Criticality

Self-organized criticality (SOC) is based upon the idea that complex behavior can develop spontaneously in certain many-body systems whose dynamics vary abruptly. Researchers have observed characteristic general behavior in systems as diverse as earthquakes, sandpiles, and even biological evolution, and have suggested SOC as a way of understanding this behavior. This book is a clear and concise introduction to the field of self-organized criticality and contains an overview of the main research results.

The author begins with an examination of what is meant by the term *self-organized criticality,* the characteristics of this type of behavior, and the systems in which it can occur. He then presents and analyzes computer models (some of which are detailed in an appendix) to describe a number of systems, and explains the different mathematical formalisms developed to understand SOC. In the final chapter he assesses the impact of this field of study, and highlights some key areas of new research.

The author assumes no previous knowledge of the field, and the book contains several exercises. It will be an ideal textbook for graduate students taking physics, engineering, or mathematical biology courses in nonlinear science or complexity.

CAMBRIDGE LECTURE NOTES IN PHYSICS

Self-Organized Criticality

Emergent Complex Behavior in Physical and Biological Systems

HENRIK JELDTOFT JENSEN

Imperial College

CAMBRIDGE
UNIVERSITY PRESS

CAMBRIDGE UNIVERSITY PRESS
Cambridge, New York, Melbourne, Madrid, Cape Town, Singapore,
São Paulo, Delhi, Dubai, Tokyo, Mexico City

Cambridge University Press
The Edinburgh Building, Cambridge CB2 8RU, UK

Published in the United States of America by
Cambridge University Press, New York

www.cambridge.org
Information on this title: www.cambridge.org/9780521483711

© Cambridge University Press 1998

First published 1998

A catalogue record for this publication is available from the British Library

Library of Congress Cataloguing in Publication Data

Jensen, Henrik Jeldtoft, 1956–
Self-organized criticality / Henrik Jeldtoft Jensen.
p.   cm. – (Cambridge lecture notes in physics ; 10)
Includes bibliographical references and index.
ISBN 0-521-48371-9 (pbk.)
1. Critical phenomena (Physics).   2. Complexity (Philosophy).
3. Physics – Philosophy.   I. Title.   II. Series.
QC173.4.C74J46   1996
003–dc21   97-10652
                                            CIP

ISBN 978-0-521-48371-1 Paperback

To Annalise and Vagn Jeldtoft Jensen

Contents

Appendices

Preface

This is a book intended for everyone interested in one of the most exciting and ambitious current developments in the field of physics and complex systems. A little bit of mathematics background may be helpful in certain sections of the book, but large parts can be read without any special prerequisites.

First we describe, in Chapters 1 and 2, what is meant by the notion *self-organized criticality* (SOC). We also list the characteristics of this behavior. In Chapter 3 we discuss a variety of systems that might exhibit the kind of behavior denoted by SOC. We discuss such systems as sandpiles, superconductors, earthquakes, and biological evolution. In Chapter 4 we describe various computer models of SOC. Most of these models are so simple that anyone with a PC can start right away to do numerical experiments on the models. Chapter 5 contains some mathematics. This chapter is dedicated to a discussion of different mathematical formalisms developed in order to understand the behavior of the computer models and so perhaps supply a mathematical description of real systems. Chapter 6 contains an attempt to assess to what extent the dream has come true – is, in fact, SOC ubiquitous? Several computer codes are included as appendices. It is hoped that this will help the interested reader to start out on the path of numerical experiments. Some mathematical details are likewise deferred to the appendices.

I believe the book can be read in at least three different ways. Those who are interested only in the overall "philosophical" impact of the ideas behind SOC can read Chapters 1, 2, and 6 for a start. Those mainly interested in the aspect of mathematical modeling via computer simulations should read Chapters 1, 2, and 4. The computer simulator might also find it inspiring to take a look at Appendices A, B, and C. The mathematician who is curious to learn about the beautiful formalisms sparked off by the development of SOC should focus on Chapter 5. It might, indeed, be useful *not* to read the book page by page. Most of the systems we discuss are considered from three different angles. With regard to earthquakes, for example, we first discuss their experimentally observed behavior. Next we present the numerical modeling developed in the context of SOC with the aim of describing the fluctuations observed in real earthquakes. Finally, we discuss various analytical developments made in the attempt to understand the numerical models. Hence, for anyone interested in a particular phenomenon or model, it would be beneficial to follow the discussion of that topic across the different chapters. This may easily be done by keeping an eye on the table of contents.

The scope of SOC is growing explosively. The very definition of the field is under debate. A huge number of scientific papers claiming to describe aspects of SOC have appeared during the last ten years. The material I have selected for the present book is obviously biased by my own outlook; another author would have followed a somewhat different path. Many topics have been left out because of space limitations. The intention was to write a short introduction to the field. The hope is that the book and its bibliography, though short and limited, can be a handy tool for those who want to break into this new and potent field.

Acknowledgment

Many thanks to Per Bak for his Toronto lecture in 1987. That started an avalanche in my own research. I thank Andrew Parry for inviting me to talk at the IOP Christmas meeting in 1993. Without that invitation there would probably have been no book. I thank Philip Meyler of Cambridge University Press for asking me to write this book.

I am indebted to D. Dhar, S. Zapperi, K. Christensen, A. Diaz-Guilera, J. V. Andersen, and N. Mousseau for helpful comments on their work. I have had the fortune to learn from collaborations with J. V. Andersen, K. Christensen, H. C. Fogedby, P. Ghaffari, H. Gillhøj, G. Grinstein, T. Hwa, O. G. Mouritsen, and S. Lise. I thank Arijeet Datta for a very efficient test-reading of the manuscript. I am grateful to the Department of Mathematics at Imperial College for facilitating a most civilized and stimulating environment. Special thanks to Steven Spencer, Jim Totty, Peyman Ghaffari, Nathan Lindop, Sefano Lise, Marianne Storey, Arijeet Datta, and David Jackson for volunteering to be my students.

As always, I am thankful to Vibeke and Barbara. Without their persistence, nothing would be much fun.

1

Introduction

Consider a collection of electrons, or a pile of sand grains, a bucket of fluid, an elastic network of springs, an ecosystem, or the community of stock-market dealers. Each of these systems consists of many components that interact through some kind of exchange of forces or information. In addition to these internal interactions, the system may be driven by some external force: an electric or a magnetic field, gravitation (in the case of sand grains), environmental changes, and so forth. The system will now evolve in time under the influence of the external driving forces and the internal interaction forces, assuming we can break the system up into internal and external components in an unproblematic way. What happens? Is there some simplifying mechanism that produces a typical behavior shared by large classes of systems, or will the behavior always depend crucially on the details of each system?

The paper by Bak, Tang, and Wiesenfeld (1987) contained the hypothesis that, indeed, systems consisting of many interacting constituents may exhibit some general characteristic behavior. The seductive claim was that, under very general conditions, dynamical systems organize themselves into a state with a complex but rather general structure. The systems are complex in the sense that no single characteristic event size exists: there is not just one time and one length scale that controls the temporal evolution of these systems. Although the dynamical response of the systems is complex, the simplifying aspect is that the statistical properties are described by simple power laws. Moreover, some of the exponents may be identical for systems that appear to be different from a microscopic perspective.

The claim by Bak, Tang, and Wiesenfeld (BTW) was that this typical behavior develops without any significant "tuning" of the system from the outside. Further, the states into which systems organize themselves have the same kind of properties exhibited by equilibrium systems at the critical point. Therefore, BTW described the behavior of these systems as *self-organized criticality* (SOC). The hope was that here was a dynamical explanation of why so many systems in nature exhibit complex spatial and temporal structures. Self-organized criticality became a candidate for the sought-after theory of complexity (Waldrop 1992). One reason for the intense interest SOC has received is that

1

it combines two fascinating concepts – self-organization and critical behavior – to explain a third, no less fascinating and fashionable, notion: complexity.†

Phenomena in very diverse fields of science have been claimed to exhibit SOC behavior. It started out with sandpiles, earthquakes, and forest fires. Next came electric breakdown, motion of magnetic flux lines in superconductors, water droplets on surfaces, dynamics of magnetic domains, and growing interfaces. The idea was soon suggested to apply to economics, and SOC models have more recently been proposed as ways of understanding biological evolution.

There does not exist a clear-cut and generally accepted definition of what SOC is. Nor does a very clear picture exist of the necessary conditions under which SOC behavior arises. Part of the purpose of the present book is to focus on these issues and contribute to a clarification of what we should, more precisely, mean by SOC. We will find that some of the experimental systems first taken as examples of SOC might not actually be so.

Connected with this lack of definition of SOC is a lack of a mathematical formalism. Despite an enormous interest from people working in many different fields, SOC still lacks a well-established mathematical framework. We are searching for a formalism equivalent to the partition function and the free energy in equilibrium statistical mechanics.

The aim of equilibrium statistical mechanics can be summarized in a very precise and simple way. It is the branch of physics concerned with the calculation of the partition function of a physical system. Define first the Hamiltonian of the considered system, that is, a function H of the configurations $\{r\}$ that determine the energy of a given configuration: $E = H[\{r\}]$. As soon as one has defined the Hamiltonian, all that equilibrium statistical mechanics need do is to perform the sum

$$Z = \sum_{\{r\}} \exp\left\{\frac{-H[\{r\}]}{k_B T}\right\}, \tag{1.1}$$

where k_B is Boltzmann's constant, T is the temperature, and the sum is over all possible configurations of the system. No general formalism like this has been established for SOC, although some individual formalisms have been developed. Exact as well as approximate schemes exist that can be applied in specific cases. The mathematical foundation of SOC is emerging.

It is useful to understand what the name *self-organized criticality* is meant to imply. The term consists of two parts. *Self-organization* has for many years been used to describe the ability by certain nonequilibrium systems to develop

† After the present volume had already been sent to the printer, I became aware of an interesting and colorful book by Per Bak, ambitiously entitled *How Nature Works* (Springer-Verlag, 1996). The book contains a nontechnical description of the historical development of, and the general ideas behind, self-organized criticality.

structures and patterns in the absence of control or manipulation by an external agent (Nicolis 1989). Examples include the growth of patterns in chemical reactions and, to make an ambitious leap, the development of structure in biological systems. The word *criticality* has a very precise meaning in equilibrium thermodynamics (Binney et al. 1992). It is used in connection with phase transitions (strictly speaking, continuous transitions). When the temperature of the system is precisely equal to the transition temperature, something extraordinary happens. For all other temperatures, one can disturb the system locally and the effect of the perturbation will influence only the local neighborhood. However, at the transition temperature, the local distortion will propagate throughout the entire system. The effect decays only algebraically rather than exponentially. Although only "nearest neighbor" members of the system interact directly, the interaction effectively reaches across the entire system. The system becomes *critical* in the sense that all members of the system influence each other.

The critical behavior of thermodynamic systems is well understood. The approach of the critical temperature can, by use of Wilson's renormalization group theory (Binney et al. 1992), be described mathematically in minute detail starting from the formalism describing the thermodynamic free energy of the system considered. The fundamental reason for the possibility of such a mathematical description is that we have Gibb's canonical ensemble. In order to make statistical considerations, one must be able to ascribe the correct probability to the various possible states of the system. This is precisely what the Boltzmann factor $\exp(-H/T)$ does. All we need to know is the energy, $H[\{\mathbf{r}\}]$, of a given configuration.

We do not know of an equivalent formalism that determines the probability with which given configurations of a system will occur during the evolution of the dynamical equation governing that system. Hence, we are not in general able to calculate the statistical properties, such as correlation functions, of dynamical systems. Despite this lack of mathematical underpinning of the statistical description of dynamical systems, BTW suggested that a large group of systems behaves very much like thermodynamical systems right at the phase transition temperature. Moreover, dynamical systems will drive *themselves* into states characterized by algebraic correlations – unlike systems in thermodynamical equilibrium, for which *tuning* is essential.

What kind of systems will evolve into a SOC dynamical state? A separation of time scales is required. The process connected with the external driving of the system needs to be much slower than the internal relaxation processes. The prototypical example is an earthquake. The stress in the earth's crust is built up on the scale of years owing to the motion of the tectonic plates. The stress is subsequently released in a few seconds or minutes during an earthquake.

The separation of time scales is intimately connected with the existence of *thresholds* and *metastability*. It is the existence of a threshold that ensures the separation of time scales. Think again about solid friction – or earthquakes. Say you want to push your piano across the floor. You slowly increase the force you apply to the piano. At first, nothing happens; the piano is stuck. The stress between the floor and the bottom of the piano builds up as the applied force increases. At a certain point, the friction forces between the floor and the piano are not able to sustain the applied force any longer. The piano does a rapid jump ahead, and the stress in the piano–floor interface is released. The applied force drops, the piano is stuck again, and the cycle starts over.

The applied force has to build up in order to overcome a certain threshold. This occurs over a time scale much longer than the short time interval it takes the piano to jump ahead. During the build-up phase, energy is gradually stored. This energy is then released, or dissipated, nearly instantaneously in the moment the piano moves forward. If no threshold for motion existed – that is, if the piano were, say, standing on ice – then the piano would move ahead continuously and the energy would be dissipated at the same rate as it is pumped into the system.

The actual friction force that the piano must overcome at a given moment will depend on the microscopic details of how the rough piano bottom interlocks with the rough surface of the floor. This means that there are many different states in which the piano will remain stuck even in the presence of an applied force. All these states are *metastable*. The friction forces induce strain in the floor as well as in the piano, and this strain corresponds to a certain amount of stored elastic energy. Thus, despite the piano–floor system being in a stable (i.e., time-independent) state, the system is not in the lowest energy state. It is in one of many metastable states.

Among all the metastable states, some are of particular importance: the set of configurations visited by the piano as it performs the jerky motion. These states are *marginally* stable. A slight increase in the applied force can lead to almost any response. Sometimes the increase in the force will be able to bring the piano forward by a small jump. At other times the increase in applied force might not even be able to make the piano move, and at still other times the same amount of increase of the driving force might induce a large jump forward.

Bak, Tang, and Wiesenfeld originally envisaged the marginally stable states as characterized by the lack of any typical time or length scale. This is precisely the case for the configurations of a thermodynamic system at the critical temperature. The lack of a typical scale leads to algebraic correlation functions.† We shall discuss at length the properties of the SOC state in several

---

† Note that a power law $f(x) = x^a$ has the property that the relative change $f(kx)/f(x) = k^a$ is independent of x. In this sense, power laws lack a characteristic scale.

models. One finds, as anticipated, that the distribution functions describing the frequency with which various events occur in the SOC state exhibit power laws. The Guthenberg–Richter law for the distribution of energy releases in earthquakes is a power law. If E is the energy released during an earthquake, then the probability for an earthquake of that size is given by $P(E) \sim E^{-B}$. This kind of distribution is seen again and again in SOC model systems. For example, in toy models of sandpiles one finds that the distribution of lifetimes of the avalanches as well as the distribution of avalanche sizes follow power law behavior.

The original ambition of the BTW paper was to explain why spatial fractals and fractal time series, known as "$1/f$ fluctuations," are so ubiquitous in nature. The properties of fractals have been studied intensively over the last one or two decades. We know how to characterize the fractal in terms of various fractal dimensions. We know how random walkers behave on fractals. The elastic properties of fractals, as well as phase transitions on fractal structures, have been studied. Despite these investigations, very little is known about why fractals are formed. What aspects of the evolution or dynamics of macroscopic systems are responsible for the formation of fractals? Many materials form crystalline structures, for example, metals and common kitchen salt. We know why this is the case: the principle of lowest energy selects the ordered crystalline phase. Fractals are certainly not the lowest energy configuration that can be selected in thermodynamic equilibrium, hence some kind of dynamical selection must occur.

How can SOC possibly be an explanation of $1/f$ noise and of fractals? The speculation by BTW was as follows. A signal will be able to evolve through the system as long as it is able to find a connected path of above-threshold regions. When the system is either driven at random or started out from a random initial state, regions that are able to transmit a signal will form some sort of random network. This network will be modified, or correlated, by the action of the internal dynamics induced by the external drive. The dynamics stop every time the internal dynamics have relaxed the system, so that all local regions are below threshold. The slow external drive will eventually bring some region above threshold once again, and the internal relaxation will restart. The result is a complicated, delicately interwoven web of regions that are coupled dynamically. When we continue to drive the system after this marginally stable SOC state has been reached, we will see flashes of action as the external perturbation manages to spark off activity through different routes of the system. The intricate nature of the combined operation of the external drive and the internal relaxation of the threshold dynamics makes it natural to imagine that the network of connected dynamical paths has some sparse percolationlike geometry.

It could well be that the structure of this dynamical network has a fractal geometry; at least this was the suggestion of BTW. If the activated regions consist of fractals of various sizes, then the duration of the induced relaxation processes traveling through these fractals can also be expected to vary greatly. It is well known that many different-acting time scales can, under certain circumstances, lead to $1/f$ noise. Bak, Tang, and Wiesenfeld imagined that this is precisely what happens in SOC systems.

All this is rather abstract, heuristic wishful thinking. In the remainder of this book we will try to make these ideas more concrete and test them on specific physical systems as well as on models.

2

Characterization of the SOC State

The nature of the critical state is described by the response of a system to external perturbation. For systems exhibiting noncritical behavior, the reaction of the system is described by a characteristic response time and characteristic length scale over which the perturbation is felt spatially. Although the response of a noncritical system may differ in detail as the system is perturbed at different positions and at different times, the distribution of responses is narrow and is well described by the average response. For a critical system, the same perturbation applied at different positions or at the same position at different times can lead to a response of any size. The average may not be a useful measure of the response; in fact, the average might not even exist.

2.1 Response Distributions

To illustrate the description of the critical state, consider the sandpile. We probe the state by adding one single grain of sand to a (randomly) chosen position on the slope. The extra grain will induce an avalanche characterized by such spatial and temporal measures as the total number s of sand grains involved in the avalanche and the lifetime t of the avalanche. We denote the statistical distributions† describing the response by $P(s)$ and $P(t)$. In the critical state we expect broad power law distributions of the form $P(s) \sim s^{-\tau}$ and $P(t) \sim t^{-\alpha}$. These distributions will typically be bounded by some lower cutoff s_1 and t_1. For example, an avalanche cannot involve the displacement of less than a single grain of sand, and the duration of an avalanche cannot be shorter than the time it takes one grain to move a distance equal to the size of a single grain. For finite systems of linear size L, a *crossover* to exponential decay often exists above a certain scale; say, $P(s) \sim \exp(-s/s_2)$ for $s > s_2$. If the system is in

---

† The mathematical literature uses the terms *density function* $p(x)$ and *distribution function* $P(x)$. The distribution function $P(x) = \int_{-\infty}^{x} p(x') \, dx'$ describes the probability that a stochastic variable X assumes a value smaller than x. The density $p(x)$ times dx is the probability that X assumes a value in an interval of length dx around x. In the physics literature, the term *distribution* is often used for either $P(x)$ or $p(x)$, often without clear specification. In the case of power laws, the exponent for $P(x)$ and $p(x)$ will differ by one, which can lead to a bit of confusion. We will use the mathematical convention whenever it is likely to clarify the discussion.

a genuine critical state then the crossover scale must be an increasing function of the system size. Often one finds that $s_2 \sim L^\omega$ with some exponent $\omega > 0$. If the exponent of the distribution is smaller than 2 (i.e., if $\tau < 2$), then the average of the distribution will not exist in the limit of infinite system size. Similarly, if the exponent is smaller than 3 then no second moment will exist, and the "width" – defined as the standard deviation – is infinite.

2.2 Temporal Fluctuations

Self-organized criticality was first introduced under the subtitle *An explanation of 1/f noise*. We shall in this section explain what is meant by this title, why it is a very ambitious title, why the claim was wrong in a very straightforward technical sense, and why – despite this flaw – there might still be something in the hypothesis.

Let us start with the term "$1/f$ noise." This is an imprecise label used to denote the nature of a certain type of time correlation. Assume that we measure some time-dependent signal $N(\tau)$, which could be sunspot activity, fluctuations in the electrical resistance of a conductor, the flow of the river Nile, pressure variations in the air caused by music, or the total activity in sandpile cellular automata (see Press 1978; Duta and Horn 1981; Weissmann 1988). Hence, many very different time signals exhibit $1/f$ fluctuations. This has been known for about a century; nevertheless, an explanation of why $1/f$ is ubiquitous is lacking. It was this long-standing puzzle the BTW paper claimed to unravel.

We now explain why $1/f$ is intriguing. Consider one of the time signals just listed. The signal fluctuates up and down in a seemingly erratic way. We wonder if the value of the signal at $N(\tau_0)$ influences what is to be measured at a later time $N(\tau_0 + \tau)$. We are not interested in any specific time instant τ_0 but rather in the typical (i.e., the statistical) properties of the fluctuating signal. The amount of dependence, history, or causality in the signal can be characterized by the temporal correlation function

$$G(\tau) = \langle N(\tau_0)N(\tau_0 + \tau)\rangle_{\tau_0} - \langle N(\tau_0)\rangle_{\tau_0}^2. \qquad (2.1)$$

If there is no statistical correlation between the signal at τ_0 and at τ time units later, we have $G(\tau) = 0$. The speed with which $G(\tau)$ decreases – from the average instantaneous fluctuation $G(0) = \langle N(\tau_0)^2\rangle_{\tau_0} - \langle N(\tau_0)\rangle_{\tau_0}^2$ toward zero – measures the length of the duration of the correlations or memory effects in the signal.

See Appendix D for a specific illustration of the relation between the power spectrum and the original time signal.

2.3 Power Spectrum and Distribution of Lifetimes

The power spectrum is defined in terms of the square amplitude of the Fourier transform of the time signal. That is,

$$S(f) = \lim_{T \to \infty} \frac{1}{2T} \left| \int_{-T}^{T} d\tau \, N(\tau) e^{2i\pi f \tau} \right|^2. \tag{2.2}$$

The correlation function is, for a stationary process, related to the power spectrum through a cosine transform as follows (see Appendix D):

$$S(f) = 2 \int_{0}^{\infty} d\tau \, G(\tau) \cos(2\pi f \tau). \tag{2.3}$$

Time correlations are therefore often discussed in terms of the power spectrum. A heuristic argument indicates the special nature of $1/f$ fluctuations. Assume that $S(f) \sim 1/f^\beta$ and that $G(\tau) \sim 1/\tau^\alpha$. Substitute $x = 2\pi f \tau$ in (2.3), and forget for the moment any worry concerning the convergence of the integral. From (2.3) it then follows that $1/f^\beta \sim 1/f^{1-\alpha}$. Thus, when β is close to unity, the exponent α must be close to zero. For β exactly equal to unity, the assumed form $G(\tau) \sim 1/\tau^\alpha$ breaks down and is replaced by a slow logarithmic decay. This means that power spectra of the form $S(f) \sim 1/f^\beta$ correspond to extremely long time correlations when $\beta \simeq 1$. This is the reason why $1/f$ fluctuations are considered to be of particular interest. See Appendix D for details on power spectra and correlation functions (see also Weissman 1988).

The 1987 BTW paper claimed that their model exhibited this kind of $1/f$ fluctuation. This turned out not to be correct (Jensen, Christensen, and Fogedby 1989) – at least not in the way believed by BTW. Nonetheless, the BTW model possesses interesting long-time correlations. We shall return to this point in Section 4.2.5.

It is tempting to try to understand the correlations within a time signal by means of a decomposition of that signal into a linear random superposition of independent components. An old example of this approach is due to van der Ziegle (1950), who showed how one can obtain power spectra of the form $1/f^\beta$ from a weighted superposition of Poisson processes. Each individual process is assumed to have a characteristic time constant $\tau_c = 1/f_c$. The autocorrelation function decays exponentially: $G_{\tau_c}(\tau) \sim \exp(-\tau/\tau_c)$. The corresponding power spectrum is $S_{f_c} \sim f_c/(f_c^2 + f^2)$. By assuming no correlations between the individual processes and a specific probability density $D(f_c)$ of the characteristic frequencies, one obtains the power spectrum for the total process:

$$S(f) = \int df_c \, \frac{f_c D(f_c)}{f_c^2 + f^2}. \tag{2.4}$$

It is now possible to construct a power spectrum with a power law behavior in a certain interval by choosing the density $D(f_c)$ appropriately (see Appendix D).

The same philosophy has been applied to SOC models. Consider the avalanches in the original BTW sandpile model. Imagine pouring sand slowly onto the pile at random positions. This will lead to avalanche activity on the slope of the pile. If the pile is large enough, one might imagine that no important interference between the individual avalanches takes place. If this is the case then the total instantaneous activity on the pile may be statistically described by considering a random linear superposition of individual avalanche signals. The power spectrum of the total activity on the pile is then calculated as the appropriately weighted sum of power spectra of the individual avalanches (Christensen, Fogedby, and Jensen 1991). See Appendix D for details.

One may rightly mistrust this approach. It is not clear how much sense it makes to develop models and theories of strongly correlated processes and then to describe them as a linear superposition of independent events. Instead of linearly superimposing the flow activity induced by individual grains, one could simulate directly the flow induced by a continuous slow sprinkling of grains onto the pile. The power spectra generated in these two ways are, however, found to be numerically indistinguishable – as long as one makes sure to sprinkle gently enough (Jensen et al. 1989).

Attempts have also been made to obtain the power spectrum of signals in SOC models from stochastic diffusion equations or Langevin equations. We will return to this point in Chapter 5, where we mention that this hydrodynamic description evidently applies to the behavior of systems at time scales that are *long* when compared to the longest lifetime of the avalanches.

2.4 Spatial Correlation Functions

In the theory of equilibrium critical phenomena, the spatial correlation function $G(r)$ plays a fundamental role. Let the system be described by a field $n(\mathbf{r}, t)$. Here one can think of $n(\mathbf{r}, t)$ as, for example, the local density of particles (if we consider a liquid) or the local magnetization (in the case of a magnetic system). The correlation function is defined as

$$G(r) = \langle n(\mathbf{r}_0 + \mathbf{r})n(\mathbf{r}_0) \rangle_{\mathbf{r}_0} - \langle n(\mathbf{r}_0) \rangle_{\mathbf{r}_0}^2, \qquad (2.5)$$

where a thermal average as well as an average over the position $\mathbf{r}_0$ is assumed.

Away from the critical temperature the correlations decay exponentially, $G(r) \sim \exp(-r/\xi)$, beyond the correlation length ξ. The correlation length diverges, $\xi \sim |T - T_c|^{-\nu}$, as the critical temperature is approached. At the critical point $T = T_c$, the correlation function changes functional behavior from exponential to algebraic dependence upon r: $G(r) \sim r^{-\eta}$. The divergence of ξ is considered to be the signal of the lack of a characteristic length scale at T_c.

In the same way, it would be obvious to describe the dynamical critical state of SOC systems by investigating their spatial correlation functions. This has not been done to any great extent because, numerically and experimentally, it is easier to obtain good statistics for the distribution functions discussed in Section 2.1. The assumption is that, if no scale shows up in the distribution functions, then neither will there be a characteristic scale in the correlation function. This is not true in complete generality. We shall later consider random neighbor models (see Sections 4.3.3 and 4.6.2), wherein a new neighborhood is assigned to the sites of the models in every time step. The repeated changes in the neighborhood make it meaningless to talk about a spatial structure of the model. In this way, one eliminates the effect of spatial correlations, which makes analytic progress more tractable. Although the spatial correlations are destroyed, in some cases the event distributions follow power laws. Needless to say, the significance of such power laws is somewhat different from those in models with a spatial structure.

3

Systems Exhibiting SOC

3.1 Introduction

In this chapter we examine the extent to which self-organized criticality is of relevance to real physical systems. A large number of experiments have purported to reveal generic SOC behavior. We do not pretend to know the final and definite answer. Some systems are more accessible to experimentation than others. It is easier to settle the question concerning the distribution of avalanche sizes in a pile of a certain granular material than to determine the properties of biological evolution. Self-organized criticality might, in the end, not be the most useful way to describe some of the dynamical systems we discuss in this book. However, I consider it an undeniable achievement the degree to which SOC developments have revived interest in the dynamics of (say) sandpiles. In this chapter we discuss the phenomenological implications of a set of experimental observations.

According to the seminal 1987 paper by Bak, Tang, and Wiesenfeld (BTW), the hallmark of SOC is its lack of any scale, in time as well as in space. As a consequence, we observe spatial fractals and temporal $1/f$ fluctuations. Thus the strategy for our experimental search for SOC is clear. Measure some of the time-dependent quantities of the system. Construct the power spectrum of the signal. If the spectrum behaves like $1/f^\beta$ with $\beta \simeq 1$, must we then be dealing with SOC? No, not necessarily so (O'Brien and Weissman 1992, 1994). In fact, most power spectra found in connection with the search for SOC extend only over a narrow frequency interval. Characteristic frequencies might be hidden below the smallest frequency measured. And even if we *are* dealing with a truly critical system, not all observables will exhibit $1/f$-like spectra. For example, assume that the power spectrum of the quantity $N(t)$ behaves like $1/f^3$. Then the spectrum of the time derivative $dN(t)/dt$ will behave like $1/f$. Also, one should keep in mind that the measured spectrum might depend crucially on how the system is driven (Jensen 1991). And finally, but not least, it is necessary to establish the existence of power laws for temporal as well as spatial quantities before one can safely conclude that a system is critical.

Although $1/f$-like spectra might be indicative of critical behavior, they do not guarantee it. There are plenty of ways to produce $1/f$ spectra without any underlying critical state (O'Brien and Weissman 1992, 1994). The idea of SOC is that $1/f$ spectra *and* fractals arise as a consequence of the absence of any characteristic scale. Hence, one must establish that a system exhibits spatial fractals in addition to $1/f$ spectra. Identifying spatial fractals is in general very difficult. Instead, most experiments measure the size distribution of response events – that is, "tickle" the system in some way and measure the duration and spatial extent of the induced response. The response often takes the form of a chain reaction or generalized avalanche.

We will discuss avalanche dynamics in both sandpiles and ricepiles as examples of experimentally accessible systems. Piles of sand were taken as the picturesque illustration of SOC in the very first paper by BTW (1987). Subsequent experimentation disagreed on the form of the size distribution of the avalanches in real sandpiles. It was suspected that inertia prevented the sandpile from exhibiting critical behavior. This suspicion lead Pierre Evesque of Ecole Central in Paris to consider the possibility of placing a sandpile in outer space. The idea was to equip one of the future flights of the European Space Agency (ESA) with a sandpile in a centrifuge. This would make it possible to study granular dynamics in gravitational fields ranging from 10^{-6} to about 10^3 times the gravitational field at sea level on earth; the role of inertial effects should then become clear. The space sandpile and ESA's space flights remain future possibilities. The role of inertia was studied in a more down-to-earth system by researchers at the University of Oslo. They decided to look at ricepiles instead of sandpiles.

A completely different type of avalanche consists of magnetic flux pulses in type-II superconductors. Beautiful experiments have been performed on the dynamics of the magnetization of superconductors. We shall explain how these experiments can be interpreted as a study of avalanches of interacting flux lines. Some experiments do indeed find behavior that is consistent with SOC.

A particularly attractive system consists of water droplets formed on a window pane. All children – and some adults – have enjoyed the drama of evolving river patterns and drop formation on a window pane hit by the rain. Controlled experiments by the Ann Arbor group find that the distributions of droplets exhibit power law behavior.

To stress the variety of the systems expected (or rather, perhaps, suspected) to exhibit SOC behavior, we shall discuss two different systems. The first one is earthquakes; the other, biological evolution. The famous Guthenberg–Richter law of earthquake sizes states that the frequency of earthquakes of a given size is an inverse power of the size (energy released) of the earthquake. Biological evolution as monitored by fossil records appears to exhibit great fluctuations. Long periods of relatively quiet development are separated by violent bursts of evolutionary activity. Obviously, these two systems are difficult to investigate experimentally, although downscaled laboratory models of earthquakes have been studied. Therefore, the discussion concerns to some extent just what the geophysical observations of earthquakes – or the pantological studies of fossils – actually show.

3.2 Sandpiles

The introduction in 1987 of BTW's SOC concept employed the language of sand avalanches. They proposed a numerical model of the most essential features of sand dynamics. (We discuss the BTW cellular automaton in Section 4.2.) Despite some flawed conclusions in the 1987 paper, the BTW cellular automaton is indeed characterized by power laws and exhibits critical behavior. The situation is different for real physical sandpiles. Soon after the publication of BTW (1987), experimentalists started to study avalanches in carefully controlled conditions on different types of sand.

The group at Chicago University (Jaeger, Liu, and Nagel 1989) performed two sets of avalanche experiments (see also Jaeger and Nagel 1992). In one experiment they studied avalanches induced by sprinkling grains onto a pile. This is the exact equivalent of the system discussed by BTW. Jaeger and associates also studied a rotating drum configuration (see Figure 3.1). They examined glass beads and aluminum oxide particles. In all cases they found a narrow distribution of avalanche lifetimes. There were no signs of power laws or critical behavior. They concluded that the typical mode of the steady-state dynamics is one in which the angle Θ of the slope of the pile oscillates between two extrema: the angle of repose Θ_r and a maximum slope $\Theta_m > \Theta_r$. As the system is driven, Θ gradually increases until it reaches the value Θ_m at which a landslide is induced involving more or less the whole slope. The avalanche persists until the slope is lowered to Θ_r. There is only a small variation in the time interval between avalanches and in their duration. The dynamics of the system resembles oscillatory behavior much more than it does critical dynamics.

Power laws have been identified in other experiments. A pile – placed on a circular base and mounted on a scale – was studied by Held et al. (1990) and later by Rosendahl, Vekić, and Kelly (1993). Grains were dropped at a very low rate (one grain every few seconds), and the induced fluctuations in the total mass of the pile were monitored. Small sandpiles were found to exhibit a certain degree of power law behavior. Avalanches containing from 3 to 80 grains were found to be distributed according to $P(s) \sim s^{-2.5}$, where s is

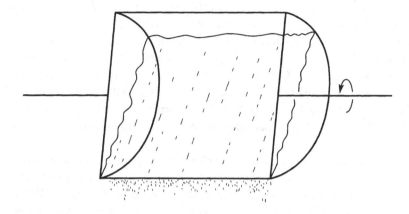

Figure 3.1. The rotating drum experiment

the mass carried out of the system by the avalanche. When Held and associates made the diameter of the base of the pile larger, they observed a crossover to an oscillatory behavior similar to that of the rotating drum system.

Rosendahl and collaborators also observed this difference between small and large piles. They found small avalanches were distributed according to a power law $P(s) \sim s^{-2.2}$, where s is the number of grains lost by the pile as an effect of the avalanche. Oscillatory behavior occurred when the pile was made larger. But in between oscillations of the large, system-spanning avalanches they observed small avalanches. These small avalanches have a tendency to be lost in the finite resolution used to consider the larger piles; the overall dynamics are controlled by the larger avalanches. Nevertheless, the small avalanches occur also in the larger piles, and they are always power-law distributed. Unfortunately, the power law of the small avalanches spans only sizes from 2 to 20 grains!

The rotating drum was revisited by Bretz et al. (1992). In this experiment, the flow down the slope was investigated (the Chicago group had analyzed only the flow over the rim of the system). Bretz and associates found behavior similar to the observations of Rosendahl et al. (1993). The evolution of the system involves large avalanches occurring at approximately evenly spaced time intervals. The large avalanches occur when the slope of the pile reaches Θ_c. The effect of these avalanches is to bring the slope back to Θ_r. However, while the average slope of the pile moves from Θ_r to Θ_c, small avalanches occur. The sizes of the small avalanches follow a power law distribution $P(s) \sim s^{-2.134}$ for about one decade of s-values. These small avalanches are typically confined to the interior of the slope's surface, and so would not be seen in a measurement of flow over the edge of the drum. An interesting finding is that the temporal behavior of the small avalanches is similar to the so-called Omori law for temporal separation of aftershocks following a large earthquake (Omori 1895). The law states that the number of earthquakes at time t after a large earthquake decays like $1/t$. The small avalanches occurring as precursors to the large avalanches are distributed in a similar way. Here the number $n(t)$ of small avalanches at time t *before* a big slide increases as $n(t) \sim 1/t$.

We conclude that these small avalanches in real sandpiles exhibit behavior that is consistent with the expectation of SOC. However, the SOC-like behavior is cut off and overwhelmed by the large avalanches in the system. Hence we must conclude that real physical sandpiles do not exhibit scale-invariant behavior in space and time.

3.3 Ricepiles

The existence of a characteristic length scale in the sandpile appears to come from inertial effects. Once an avalanche has started, it gains kinetic energy. The rolling grains become increasingly harder to stop and they eventually end up affecting the whole slope of the pile. The mass density of a rice grain is smaller than the mass density of a sand grain. One could therefore expect inertial effects to be of less significance to rice-grain than to sand-grain dynamics.

An experiment on three types of rice was carried out by a group in Oslo (Frette et al. 1996). The rice differed in aspect ratio and surface quality. Type-A rice was unpolished and elongated, with an aspect ratio (length divided by width) of 3.8. Type B consisted of polished rice with an aspect ratio of 2.0. Finally, type-C rice was polished and elongated, with an aspect ratio of 2.6. A pile of rice grains was built up between two parallel glass plates (see Figure 3.2). The separation of the two glass plates was, for all rice types, fixed at about 80% of the length of the rice grains. The grains were fed in at a slow rate at the left closed edge. Gradually, a ricepile built up between the two plates. Grains could leave the system at the right edge.

The elongated grains (type A and C) behave in a similar way, but the short and round grains of type B behave in a distinctively different manner. A large aspect ratio allows the type-A and type-C grains to build up very steep local slopes. The profile of the pile becomes very rugged, with large fluctuations in the local slope. When the grains move, they predominately slide along the surface of the pile as a more-or-less coherent cluster of grains. The dynamics of these grains are dominated by local friction.

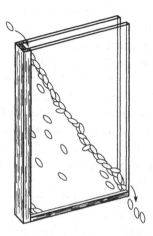

Figure 3.2. The ricepile experiment.

The shorter grains of type B move in a less collective manner than type A and C. Their motion consists mainly of individual grains rolling down the slope, either as separate single grains or as many grains moving in a fluidlike way. These grains are difficult to bring to a stop. They would typically move a distance of about half of the system size, regardless of the actual size of the system considered.

The redistribution of grains caused by each avalanche was measured. From the redistribution of mass, Frette and collaborators calculated the gravitational potential energy released by the movement of grains to lower elevation. The distribution of avalanche sizes measured in terms of released energy was determined. From a finite size scaling analysis, the avalanche distribution of type-A and type-C grains was found to follow a power law $P(E) \sim E^{-\alpha}$ with $\alpha \simeq$ 2.04. In the limit of small avalanches, the distribution $P(E)$ becomes constant. The power law was observed to apply for about $1\frac{1}{2}$ decades on the energy scale. No change in the behavior of $P(E)$ was found for increasing system size. So, contrary to the sandpiles, the power law distributions found for type-A and type-C ricepiles persisted as the size of the piles was increased.

Rice of the more circular shape (type B) is characterized by an avalanche distribution of a form like that of a stretched exponential:

$$P(E) \sim \exp(-(E/E_0)^\gamma), \tag{3.1}$$

where $\gamma \simeq 0.45$ and the characteristic avalanche size is given by $E_0 \simeq 0.45L$ (in appropriate units; see Frette et al. 1996). Here L denotes the length of the base of the pile. The increase in E_0 with increasing L suggests that there is no intrinsic energy scale of the dynamics, even for type-B rice. The probability density in (3.1) approaches a uniform density when $L \rightarrow \infty$. In this sense it seems fair to conclude that rice of type B also behaves according to the SOC scenario, but the authors conclude that type-B rice does *not* exhibit SOC behavior (Frette et al. 1996). They did this because the motion of type-B rice was observed (as described here) to be very different from the way that rice of type A and C move. Moreover, they were unable to observe a size-dependent cutoff in the power law for the avalanche size distribution with types A and C, whereas the finite size of the system always entered through the scale of the stretched exponential for type-B rice.

In conclusion, the elongated grains exhibit behavior consistent with the SOC scenario – power laws and lack of scale. The shorter rounded grains *are* characterized by a scale, so grains of type B exhibit no critical behavior. This is intuitively consistent with the idea that SOC is caused by the existence of local thresholds that lead to a large number of metastable configurations. The ability of the rounded grains to roll or flow eliminates to a large extent the threshold dynamics and greatly limits the number of possible metastable configurations for a given number of grains. The rolling grain also accumulates kinetic energy, making inertial effects relevant as in the sandpile.

3.4 Superconducting Avalanches

In this section we turn our attention to a very different type of "heap." In a spectacular experiment, Field, Witt, and Nori (1995) demonstrated that the motion of vortices in superconductors may exhibit scale-invariant behavior in both space and time. We need to go through a few details of the physics of superconductors in order to understand the experiment.

Below a certain temperature T_c, the electrons in many materials undergo a phase transition. The electrons bind together in pairs, called *Cooper pairs*. Electrons, being half spin particles, are fermions. The spin of the pair is the sum of the components and is therefore an integer. Integer spin particles are *bosons*, which is more or less the case for Cooper pairs. Bosons have a strong tendency to move in a coherent fashion.

The coherent motion of the Cooper pairs is the origin of superconductivity. In the superconducting state, electrons move without dissipating energy. Moreover, the superconducting electrons are able to screen an external magnetic field from the bulk of the sample. This is possible only for fields weaker than a certain critical value. The behavior of superconductors in stronger magnetic fields falls into two types. Type-I superconductors simply become normal conductors

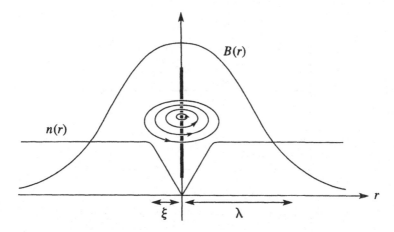

Figure 3.3. Variation in the density of superconducting electrons $n(r)$ and in the local magnetic induction $B(r)$ in the vicinity of a vortex line. The change in $n(r)$ occurs over the length scale ξ, the superconducting correlation length. The change in $B(r)$ occurs over the penetration length λ.

when the external field becomes larger than a field H_c. Type-II superconductors remain superconducting for fields H in a range $H_{c_1} < H < H_{c_2}$; for fields between H_{c_1} and H_{c_2}, the superconductor is said to be in the "mixed" state. In such a regime, the external magnetic field penetrates into the bulk of the superconductor in a nonuniform pattern. The magnetic field is concentrated along *vortex lines* consisting of circulating superconducting electrons (Tinkham 1975); see Figure 3.3.

At the core of the vortex line, the density of superconducting electrons goes to zero and the local magnetic field has a maximum. Each vortex line (or *flux line*, as they are also called) carries one quantum of magnetic flux $\Phi_0 = h/2e$. Here h denotes Planck's constant, and e is the charge of an electron. The vortex lines move closer together when the local magnetic induction increases. This leads to an increase in the magnetic field energy. From this we see that the vortex lines are mutually repulsive.

The density of superconducting electrons is lost at the core of the flux line. This leads to an interaction between the vortex core and any inhomogeneity in the superconducting material. Assume that the ability to superconduct is locally destroyed by some defect in the material. We decrease the volume in which the density of superconducting electrons is depleted when we position a flux line on top of the defective region (see Figure 3.4). The configuration with the largest volume of superconducting electrons is the most favorable one; it

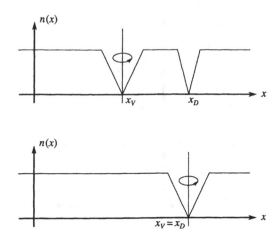

Figure 3.4. Interaction between a vortex line and a spatial inhomogeneity.

has the least free energy. This explains why there can be an attraction between inhomogeneities in the superconducting matrix and the vortex lines.

The motion of vortex lines through the superconductor is hindered by the inhomogeneities of the material, which act as "pinning centers." When the external magnetic field is increased above H_{c_1}, vortices will be created at the surface of the superconductor. The mutual repulsion between the vortices will force the vortices toward the center of the sample, yet the pinning centers will hinder this motion. Vortices will pile up on the pinning centers, and a gradient in the vortex density will be established. The forces acting on a pinned vortex from the surrounding vortices will not add up to zero when the pinned vortex finds itself in a vortex–density gradient. As the external field is increased, the vortices inside the superconductor will set up a density profile defined by the balance between the pinning forces and the forces produced by the vortex–density gradient.

Field et al. (1995) studied the motion of vortices through a superconducting material; their experimental set-up is depicted in Figure 3.5. A superconducting NbTi tube was placed in a magnetic field exterior to the tube. A pickup coil was mounted inside the bore of the tube. The external field is ramped up at a constant rate, leading to vortices being pushed into the wall of the superconducting tube. Eventually the vortices move all the way through the wall and start to enter the coil inside the tube. According to Faraday's induction law, the increasing flux inside the coil will induce a voltage V in the coil: $V(t) \sim d\Phi/dt$, where $\Phi(t)$ denotes the time-dependent flux through the coil. The flux

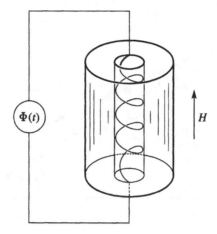

Figure 3.5. The experiment on vortex avalanches.

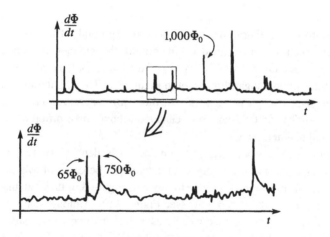

Figure 3.6. Time dependence of $d\Phi/dt$.
(Sketch of data presented in Field, Witt, and Nori 1995.)

is obtained from the measured time dependence of the voltage. Because each vortex carries one quantum of flux, the number of vortices entering the coil at a given instant is proportional to $d\Phi/dt$.

The motion of the vortices through the walls of the superconducting tube consists of a smooth continuous flow plus well-defined avalanches of vortices. The observed temporal dependence of $d\Phi/dt$ is sketched in Figure 3.6, which

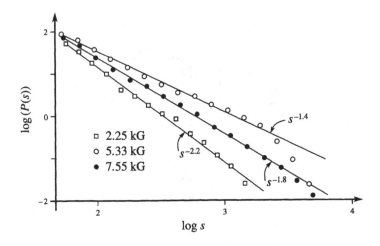

Figure 3.7. Avalanche size probability density.
(Sketch of data presented in Field et al. 1995.)

shows sharp pulses rising over the smooth background. The distribution of vortex avalanches was obtained by integrating the number of vortices contained in the individual pulses. The probability density of the avalanche sizes is sketched in a double logarithmic plot in Figure 3.7. The distribution follows a power law over a couple of decades. The three different sets of data correspond to three 450-G wide field windows centered around three different values of the external magnetic field.

To supplement the spatial analysis contained in the density of avalanche sizes, Field and associates also investigated the temporal behavior of avalanche dynamics. From time signals similar to those in Figure 3.6, they calculated the power spectrum of the vortex flow (see Figure 3.8). Although the spectra do not follow a power law over six decades as in measurements of resistive fluctuations (Weissman 1988), they do behave like $1/f^{3/2}$ for about two decades of the frequency axis.

The experiment by Field and associates shows that the stick-slip depinning of vortices in superconductors can exhibit the characteristics anticipated for SOC phenomena. However, not all measurements of vortex avalanches show this behavior. In an experiment by Zieve et al. (1996) performed on a very clean untwinned YBaCuO single crystal, the avalanche probability density was found to be strongly peaked. Simulations suggest that the difference between these experiments may be caused by the morphology of the pinning centers contained in the two materials (Pla, Wilkin, and Jensen 1996). Numerically,

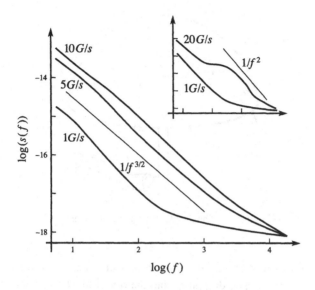

Figure 3.8. Power spectrum of the vortex avalanche signal.
(Sketch of data presented in Field et al. 1995.)

a high density of short-range pinning centers leads to a broad distribution of avalanche sizes. Longer-ranged and smoother pinning centers of lower density produce a narrow, peaked distribution.

3.5 Droplet Formation

The avalanches of all the systems considered so far consist of collections of well-defined single quantities: sand grains, rice grains, or quantized superconducting vortices. In this section we consider a different type of system, one in which the individual elements coalesce while forming the avalanche. The system consists of a dome of glass. A sprinkler sprays water droplets onto the inside of the dome, and a network of "rivers" is formed as water droplets merge and flow toward the rim of the dome. Power law distributions were found in an experiment by Plourde, Nori, and Bretz (1993).

The experimental set-up is sketched in Figure 3.9. Water is sent through a spray mister up into the inside of a Plexiglas dome, and mist condenses on the inner side of the dome. When drops acquire enough mass to overcome the sticking force of the water–glass interface, they begin to run down the surface of the Plexiglas. A moving drop will coalesce with other drops as it moves down. The

Figure 3.9. The droplet formation experiment.

drop will grow by this "snowballing" effect and an avalanche is formed as a result. The amount of water dropping from the rim of the dome is monitored as a function of time. Obviously, more than a single stream or avalanche may be operating at the same time. However, in the limit of slow deposition of mist on the dome, this problem is of less importance. The dripping from the edge was analyzed in terms of pulses of differing duration and mass. These pulses were identified as the avalanches in this system. The duration of a pulse is called the *lifetime* and the mass contained in such a pulse is the *size* of the avalanche. The influence of the viscosity of the water was investigated by running the experiment at two different temperatures. This produced a change in the viscosity by a factor of nearly 2.

The measured distribution of avalanche sizes and lifetimes exhibits power law–like behavior for about one decade. This very limited range might be due to the small system size, which for experimental reasons could not be made much larger. The higher viscosity at low temperature had a tendency to lead to large events. In this case, the viscosity allowed large clusters of droplets to form on the surface before they released an avalanche. These large clusters sweep a significant fraction of the surface. The swept surface is devoid of stationary droplets and has in this sense been reset. This resetting reminds us of the rotating drum experiment discussed in Section 3.2. The large avalanches occur only at low temperature. At higher temperatures, the drops start to run down the surface before forming clusters that are large enough to cause resetting avalanches. The dependence on viscosity also shows up in the distributions of size and lifetime of the avalanche. The low-temperature, high-viscosity system has a tendency to show a slightly broader range of sizes and lifetimes.

The power spectrum of the dripping from the rim of the dome was also measured. Unfortunately, the dynamical range was too limited to allow a convincing identification of regions of power law dependence. Indications of critical behavior were observed at low driving rate, but disappeared when the mist was sprayed on at a high rate. The distributions of sizes and lifetimes both changed to an exponential form at the high spray rate.

Water drop formation was also studied by Jánosi and Horváth (1989) in an experiment concerned with the behavior of water coverage on a tilted glass pane. After spraying water onto the pane, image processing was used to determine the distribution of droplet sizes as well as the temporal variation of total area covered by the water. The droplet distribution exhibited a clear peak around a characteristic size. The power spectrum of the temporal variation in the area covered by water behaves like $1/f^2$ and not like the magical $1/f$. A characteristic frequency f_c is expected to exist. For frequencies smaller than f_c, the $1/f^2$ behavior will cross over to a behavior like $1/f^{-\beta}$ with $\beta < 1$ (owing to normalization constraints). The authors estimate that the frequency f_c corresponds to about 39 hours. This is a very long time compared to the approximately 15-minute duration of the experiments. If the identification of this time scale is correct then – despite being noncritical – the dynamics of the water droplets in this experiment certainly exhibit extreme long-time correlations.

3.6 Earthquakes

Earthquakes occur as a result of the relative motion of tectonic plates at the outermost crust of the earth. The plates join together along faults. The plates on either side of the fault are subject to relative motion. Friction between the plates prevents the relative motion from occurring in a smooth manner. The adjacent plates stick together until stress in the interface builds up to a level that cannot be supported by the sticking power, or solid friction, between the two plates. The stress that has built up gradually over years is released within seconds or minutes during a catastrophic breakdown of the bonds that had restricted the relative motion of the plates (Scholz 1990, 1991; Sornette 1991).

We expect intuitively that this phenomenon might exhibit SOC behavior. The system is driven slowly. The driving of the system consists of the relative motion of the tectonic plates, which is characterized by a scale of a few centimeters per year. Moreover, the motion involves the key ingredient in SOC – namely, thresholds. The plates slip with respect to each other only when the local stress becomes larger than the local solid friction. The expectation that earthquakes are SOC systems seems to be met by centuries of observation of the phenomenology of earthquakes (Scholz 1990, 1991; Sornette 1991).

The size of an earthquake is measured in terms of the amount of energy E released when the plates slip. The distribution of earthquake sizes, $P(E)$, is found to be very broad indeed. A significant part of the distribution is found to follow a power law behavior $P(E) \sim E^{-B}$, where the exponent B exhibits some geographical dependence and is found to be in the interval from about 1.8 to 2.2. The morphology of the faults exhibits features characteristic of fractals. And finally, the temporal frequency of aftershocks following in the wake of a major earthquake decays with a power law known as the Omori law (Omori 1895). Let $n(t)$ denote the number of aftershocks occurring at time t after a major earthquake; observations show that $n(t) \sim t^{-A}$, with A varying from about 1 to 1.5.

Earthquakes certainly exhibit a broad range of responses, and their spatial and temporal behavior may best be described in terms of fractals and power laws. However, there appear to exist characteristic scales for earthquakes (Scholz 1990, 1991). One scale is set by the thickness of the layer of the earth's crust, the *schizosphere,* that is ruptured by the earthquake. The schizosphere has a certain thickness reaching from the earth's surface down to the plastosphere. Small earthquakes can propagate completely within the schizosphere without reaching the boundary, whereas large earthquakes are constrained by the finite thickness of the schizosphere. Small earthquakes propagate in the two-dimensional plane of the fault; large earthquakes can only propagate along the fault. Both earthquake types follow power law distributions for their sizes. However, the exponent of the power law is different for the two types.

In addition to the scale set by the thickness of the schizosphere, some geophysicists consider yet another scale: some large-segment length of the fault that is characteristic of its structure. They talk about the characteristic large earthquake that is specific to a certain segment of the fault. This characteristic earthquake resets the fault in the sense that it relaxes entirely the stress over the whole segment. The characteristic earthquake is presumed to re-occur at fairly well-defined intervals. The notion of characteristic large-event earthquakes has been met with suspicion by some physicists (Bak 1991). They point out that the statistics of the large earthquake are very limited because the frequency of earthquakes decays with a power law of their size. Hence, they claim that there is no observational justification for a separate category of large earthquakes. We are not going to settle the issue here.

I mention the controversy for two reasons. First, if a large-event scale does exist in earthquakes then it is reminiscent of the situation discussed previously in connection with sandpiles and the dynamics of water droplets on glass panes. Second, the large-event controversy makes us remember the following – somewhat banal, but nevertheless important – fact: a concept such as SOC may be

able to capture some aspects of a phenomenon without necessarily being able
to explain all its details. The broad power law response may very well be relevant up to a certain scale, whereupon other system-specific mechanisms set
in. Small avalanches are possibly power-law distributed owing to SOC behavior, but this ceases to be the case when inertia becomes important for the larger,
heavier sandslides and likewise for earthquakes. Large-scale inhomogeneities
in the fault structure can cut off the SOC behavior at a certain spatial extent of
the earthquake.

3.7 Evolution

Does the biological evolution of species take place gradually at a slow, smooth
pace, or does evolution occur via periods of hectic activity separated by long
intervals of tranquility? Is the total collection of species moving toward some
terminal configuration of optimal fitness that is inert to further improvement
by evolutionary changes? The rather sparse observational record of the evolution and extinction of species makes both questions impossible to answer with
certainty. However, some twenty years ago Gould (1977) suggested that evolution does take place through bursts of activity separated by calm periods.
Later, Raup (1986) presented paleontological evidence that extinction is, as he
phrased it, "episodic" at all scales. This means that species survive for long periods but then disappear within a relatively short span of years. Moreover, one
often observes that the extinction of one species occurs simultaneously with the
extinction of a number of other species. The number of extinct species during
a certain burst of extinction varies greatly. In addition to presenting evidence
for this avalanche-like nature of extinction, Raup also suggested that extinction might be a rather arbitrary process – a matter of "chance susceptibility"
to surrounding conditions, which a species is subject to at any given moment
in history. Raup offers the example of mammals and dinosaurs coexisting for
more than 100 million years. Only after the dinosaurs disappeared did mammals undergo their explosive evolutionary success, which was made possible
by the absence of dinosaurs. Dinosaurs may not necessarily have become ob-
solete in the Darwinian sense; their extinction might just as well have been ef-
fected by an unusual fluctuation in the ever-present environmental stress. Extinction, argues Raup, might not be the constructive drive behind evolution, but
rather more accidental and arbitrary than hitherto assumed.

When evolution and extinction are portrayed in this fashion, we immediately
spot similarities with the other examples of SOC dynamics discussed in this
chapter. Most of the time, barely anything happens. Then, all of a sudden, the
extinction of a certain species tips the balance in an essential way, leading to

a substantial change in the fitness of *other* species that had depended on the now-extinct species, and a chain of events or an avalanche is initiated. The extinction chain and the adjustment induced in the fitness of the interrelated species continues until a new relative balance or equilibrium has been reached. Whether or not this is the way evolution actually takes place is infinitely harder to check than whether or not piles of sand or rice relax according to the SOC description. We will not be able to settle the issue here to any degree. Nevertheless, we mention evolution as an example of one, possibly far-reaching, application of the SOC approach. In Section 4.6.2 we discuss a very simple and charming mathematical model developed by Bak and Sneppen (1993) that can be thought of as a model of evolution if evolution's dynamical laws are as depicted here.

4

Computer Models

4.1 Introduction

The present chapter is about models. All the models are inspired by some physical system: sandpiles, earthquakes, magnetic vortex motion, forest fires, interface growth, or biological evolution. The models are defined in terms of a dynamical variable – for example, the local slope of the sand heap or the stress in the earthquake fault. The dynamical variable or field is updated in every time step according to some algorithm. The choice of the updating algorithm is, to some degree, arbitrary. The criteria for choosing the relevant definitions are, for the most part, simplicity and intuition. Statistical mechanicians have this overall belief that complexity arises from simplicity: that the intricate and complex behavior found in many systems is due to the large number of degrees of freedom, rather than caused by some very complicated behavior of the individual degrees of freedom. All the models described in this chapter are formulated according to this paradigm. In addition, they are all designed with an eye toward numerical ease and efficiency when simulated on computers.

For ease of presentation, for each model we shall first simply present the definition of the model and the conclusions derived from numerical studies thereof. Since the aim is to understand the world around us, we shall use separate sections to discuss the relevance and shortcomings of each model. As always when building models of Nature, one proceeds through steps of increasing refinement, and so also for models of SOC systems. Perhaps it is worth mentioning that just because a model doesn't capture *all* features of a specific system, this need not imply that the model doesn't capture *any* aspects at all.

4.2 Sandpiles: Conservative Model

In the 1987 paper by Bak, Tang, and Wiesenfeld (BTW), a simple cellular automaton was introduced as a means of illustrating numerically the more philosophical arguments put forward concerning the typical trend of systems to organize into a critical state (see also Bak, Tang, and Wiesenfeld 1988; Wiesenfeld, Tang, and Bak 1989). In this section we shall examine features of the BTW *sandpile cellular automaton.* The reason for including the word "sandpile" in the name of the model is purely heuristic. The rules defining the dynamics of the model are derived from their intuitively apparent relation to the dynamics of sand grains tumbling down the slope of a sand heap. As discussed previously in Section 3.2, the degree of criticality observed in real piles of granular material has been much debated by experimentalists.

The idea behind the dynamics of the model is that the local gradient of the sandpile determines the stability of the pile. If the gradient $z(\mathbf{r})$ at a specific point $\mathbf{r}$ is smaller than some threshold value z_c then the sandpile is locally stable at position $\mathbf{r}$, whereas when $z(\mathbf{r})$ becomes larger than z_c the pile becomes unstable and grains will tumble down the slope of the pile. The variable z is defined on a d-dimensional square lattice and is treated as an integer-valued scalar. Because $z(\mathbf{r})$ is treated as a scalar in the simulations of the model, one neglects the fact that the gradient of the slope of the sandpile is a vector in dimensions $d > 1$. The behavior of the model is not altered significantly when $z(\mathbf{r})$ is treated as a vector (see McNamara and Wiesenfeld 1990).

4.2.1 One-Dimensional Sandpile

Let us first consider one dimension to make clear the ideas behind the choice of the dynamical rules defining the model. Consider Figure 4.1. The integer h_i denotes the number of grains in the sand column above position i. The slope will be measured by $z_i = h_{i+1} - h_i$. The dynamics of the model is defined in terms of the following two operations: (a) adding a grain to the pile, and (b) relaxing the slope of the pile whenever the local gradient exceeds the stability threshold z_c.

(a) Suppose we add a grain to position i, that is, $h_i \rightarrow h_i + 1$. The values of the z-variable will change accordingly:

$$\begin{aligned} z_{i-1} &\rightarrow z_{i-1} + 1, \\ z_i &\rightarrow z_i - 1. \end{aligned} \tag{4.1}$$

(b) Assume that $z_i > z_c$. Sand will now slide locally. A grain is taken from position i and added to position $i + 1$. This is represented by the operation $h_i \rightarrow h_i - 1$ and $h_{i+1} + 1$. The resulting change in the z-variables may be listed as

$$\begin{aligned} z_{i-1} &\rightarrow z_{i-1} + 1, \\ z_i &\rightarrow z_i - 2, \\ z_{i+1} &\rightarrow z_{i+1} + 1. \end{aligned} \tag{4.2}$$

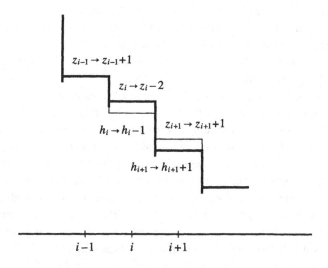

Figure 4.1. The one-dimensional BTW sandpile cellular automaton.

Note that both operations (a) and (b) leave the total z-value of the system unchanged. The integrated z-value of the system can change at the boundary.
Imagine placing a wall at the left edge of the system. The slope at the wall is
kept equal to zero; that is,

$$z_0 = 0. \tag{4.3}$$

The right-hand edge of the system is open. Sand grains can leave the system at
this edge, and this is represented by the following algorithm for updating site
N whenever $z_N > z_c$:

$$z_{N-1} \rightarrow z_{N-1} + 1,$$
$$z_N \rightarrow z_N - 1. \tag{4.4}$$

One should not worry too much about how appropriate the choice of the dynamical rules of the model is in order to represent an actual heap of granular material realistically. Instead, consider the connection to the sandpile as a
mnemonic device and a source of inspiration.

It is often more convenient to drive the model by a rule other than that defined
in (a). We might instead use the following.

(a′) Addition of sand:

$$z_i \rightarrow z_i + 1. \tag{4.5}$$

In sandpile language, this corresponds to adding a layer of sand of unit thickness for all positions to the left of position i. Perhaps one would rather aban-
don the interpretation of z as a slope completely (which will have to be done in

$d > 1$ anyway) and simply think of z_i as some unspecified dynamical variable. When the system is perturbed according to (a'), the model is often referred to as the "height" model in contrast to the "slope" model.

The behavior of the one-dimensional version of this model is easy to predict. Imagine that we start from a very steep surface: $z_i \geq z_c$ for $0 \leq i \leq N$. We let the system relax according to rule (b). The system gradually evolves toward the configuration where $z_i = z_c$ for all sites i. This state is a *global attractor;* we would end up with this state by applying successively rule (a) and (b) no matter what configuration the system started from.† This is a (perhaps trivial) example of how a dynamical system selects a very atypical configuration from the set of all possible stable configurations – a situation very different from the microcanonical ensemble of equilibrium thermodynamics in which all microscopic states with the same energy are equally likely to occur. The observation that dynamical systems tend to select atypical states played an important part in the work that led to the development of SOC (see Tang et al. 1987).

The configuration $z_i = z_c$ for all i is special in two respects. Any further addition of sand will lead to relaxation. This is an example of a *minimally stable* configuration. In addition to being minimally stable, the state is *critical* in the sense that perturbations – induced by the application of rule (a) or (a') at some randomly chosen position – will lead to events of all sizes. In the one-dimensional case, the event distribution is readily calculated.

Imagine that the pile is perturbed at some randomly chosen position. This might make the local slope *over*critical and thereby induce a relaxation. Let $P(s)$ denote the probability that the number of sites involved in the relaxation is equal to s. Let $P(t)$ denote the probability that the number of time steps needed to reestablish a stable configuration is equal to t. We notice that application of rule (a) to a site i in the minimal state will lead to an "avalanche" consisting of a single grain of sand being transported from the site i to the open end of the system at position N. The system becomes stable when the grain has tumbled over the edge of the system. A perturbation at position i involves $N - i + 1$ sites and lasts for the same number of time steps. We assume that our perturbation of the system hit all sites with equal probability. Hence, the two probability densities are identical, $P(s) = P(t)$, and equal to the uniform distribution on the set $\{1, 2, \ldots, N + 1\}$:

$$P(t) = \begin{cases} 1/(N+1) & \text{if } i \in \{1, 2, \ldots, N+1\}, \\ 0 & \text{otherwise.} \end{cases} \tag{4.6}$$

† Note that by starting from a flat surface ($h_i = 0$ for all i), one will, somewhat artificially, go through configurations with large negative values of $h_i - h_{i+1}$ simply because rule (b) does not involve the absolute value of z_i.

The one-dimensional version of the sandpile cellular automata is very easy to understand because the critical state of the system consists of a single configuration – namely, the minimally stable configuration. This is not the case when the model is generalized to higher dimensions.

4.2.2 Dimensions Larger than 1

We now generalize the dynamical rules (a), (a'), and (b) of the model to d dimensions. We define a dynamical variable $z(\mathbf{r})$ on a cubic d-dimensional lattice of linear size L. The basis vectors of the lattice are denoted by $\mathbf{e}_i$ with $i = 1, 2, \ldots, d$. There are four "operation rules" as follows.

R1 *Conservative Perturbation*

$$z(\mathbf{r}) \rightarrow z(\mathbf{r}) + d,$$
$$z(\mathbf{r} - \mathbf{e}_i) \rightarrow z(\mathbf{r} - \mathbf{e}_i) - 1 \quad \text{for } i = 1, \ldots, d. \tag{4.7}$$

R2 *Nonconservative Perturbation*

$$z(\mathbf{r}) \rightarrow z(\mathbf{r}) + 1. \tag{4.8}$$

R3 *Relaxation*

If $z(\mathbf{r}) > z_c$, then

$$z(\mathbf{r}) \rightarrow z(\mathbf{r}) - 2d,$$
$$z(\mathbf{r} \pm \mathbf{e}_i) \rightarrow z(\mathbf{r} \pm \mathbf{e}_i) + 1 \quad \text{for } i = 1, \ldots, d. \tag{4.9}$$

R4 *Boundary Conditions*

We visualize sand piling up against a corner (see Figure 4.2). The sand is able to drop off over the edges opposite the corner of the cube.

Closed Boundary. The dynamical variable is kept equal to zero along the one corner of the cube – that is, for positions $\mathbf{r} = (r_1, r_2, \ldots, r_d)$ for which at least one of the coordinates r_i is equal to zero. In other words, we have

$$z(\mathbf{r}) = 0 \quad \text{if there exists a coordinate } r_i = 0. \tag{4.10}$$

Open Boundary. Sand leaves over the edges defined by having one of the coordinates r_i equal to L. If $z(\mathbf{r}) > z_c$ for $\mathbf{r}$ along this corner, then we perform the following update:

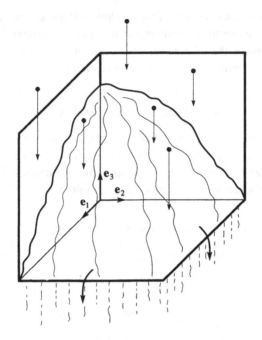

Figure 4.2. Sandpile in a corner.

$$z(\mathbf{r}) \rightarrow z(\mathbf{r}) - 2d + \text{number of } i \text{ with } r_i = L,$$

$$z(\mathbf{r} + \mathbf{e}_i) \rightarrow z(\mathbf{r} + \mathbf{e}_i) + 1 \quad \text{if } r_i \neq L, \tag{4.11}$$

$$z(\mathbf{r} - \mathbf{e}_i) \rightarrow z(\mathbf{r} - \mathbf{e}_i) + 1 \quad \text{for } i = 1, \dots, d,$$

and

$$z(\mathbf{r}) = 0 \quad \text{if there exists a coordinate } r_i = 0. \tag{4.12}$$

The lattice is updated simultaneously. If more than one site becomes over-critical, $z(\mathbf{r}) > z_c$, we update all these sites at the same time. A *time step* is defined as a visit to all the L^d sites, checking whether or not their z-value is larger than z_c and performing an update whenever $z(\mathbf{r}) > z_c$. The pile is perturbed by rule R1 or R2. If overcritical sites are produced then relaxation is performed according to rules R3 and R4 until all sites are again subcritical.

A straightforward implementation of these rules as a computer code is given in Appendix A. As soon as one has written the computer program, one can investigate the statistical properties of the model simply by letting the computer perform the basic dynamical rules over and over again. As in the one-dimensional case, the statistical properties of the model after many time steps

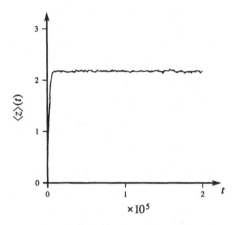

Figure 4.3. Temporal evolution of the average slope $\langle z \rangle$.
(Sketch of data presented in Christensen, Fogedby, and Jensen 1991.)

do not depend on the initial configuration. One way to probe if a statistical
stationary state has been reached is to measure the average z-value $\langle z \rangle(t)$ as a
function of time (see Figure 4.3), where the average is over all lattice sites $\mathbf{r}$.
When one starts from a configuration where all sites have z-values larger than
z_c, the average $\langle z \rangle(t)$ will initially decrease. After a certain time period, $\langle z \rangle(t)$
starts to fluctuate about an average value given by

$$\langle \bar{z} \rangle = \lim_{T \to \infty} \frac{1}{T} \int_0^T dt \, \langle z \rangle(t). \tag{4.13}$$

Starting from a flat surface with all $z(\mathbf{r}) = 0$, the spatial average $\langle z \rangle(t)$ will increase toward the same asymptotic value arrived at when starting from an over-
critical configuration. Any randomly chosen initial configuration will lead –
after a long time – to the same statistical properties. This long-time behavior
of the system is characterized by the statistical properties of the asymptotic attractor (see Figure 4.4). The attractor consists of a subset $\mathcal{A}$ of configurations.
The set $\mathcal{A}$ of configurations replaces the single marginally stable configuration
encountered in the one-dimensional version. The set $\mathcal{A}$ contains many configurations that are not marginally stable. There exist positions $\mathbf{r}$ where sand can
be added without making $z(\mathbf{r})$ overcritical. Although the configurations are
not marginally stable, the system does exhibit critical behavior as it is driven
through the configurations contained in the set $\mathcal{A}$. The next section explains
what we mean by this.

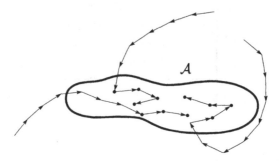

Figure 4.4. The attractor $\mathcal{A}$ consisting of the marginally stable configurations.

4.2.3 Critical Response

The dynamical properties of the cellular automata are studied in the following way. The pile is built up by successive use of one of the two perturbation mechanisms R1 or R2 followed by the relaxation rules R3 and R4 (see Section 4.2.2). When the overall drift in the average slope $\langle z \rangle (t)$ is replaced by fluctuations about the temporal mean value, one can start to collect statistics (see Figure 4.3).†

In the stationary state, the distributions of responses are measured simply by applying the perturbation and then measuring the spatial size, the total "energy release," and the temporal duration of the avalanche induced by the perturbation. Histograms can be produced simply by repeating this procedure over and over again. Before discussing the distributions, we need to define precisely what is meant by size, energy release and duration.

Choose a lattice site $\mathbf{r}_0$ at random. Perform a perturbation at position $\mathbf{r}_0$. Assume that an avalanche is induced. The avalanche will span a connected region A of $|A|$ different lattice points $\mathbf{r}$. The center of mass of the avalanche is defined as

$$\mathbf{R}_{cm} = \frac{1}{|A|} \sum_{\mathbf{r} \in A} \mathbf{r}. \tag{4.14}$$

As a measure of the *linear size* l of the avalanche we use

$$l = \frac{1}{|A|} \sum_{\mathbf{r} \in A} |\mathbf{r} - \mathbf{R}_{cm}| \tag{4.15}$$

---

† Obviously, the temporal behavior of $\langle z \rangle (t)$ offers only a rough check on whether the asymptotic regime has been reached. To be careful, one would check the temporal evolution of the distribution functions.

(Diaz-Guilera 1992). The *total energy release s* is simply defined as the total number of sites that must be relaxed before all sites again become undercritical. Notice that a specific lattice position **r** can count more than once. It may become overcritical several times as a result of neighboring sites becoming overcritical. The *lifetime t* of an avalanche is defined as the total number of simultaneous updates of the whole lattice that must be executed in order to make all sites undercritical.

The three quantities l, s, t are not independent. Hence, in principle one must measure the simultaneous probability density $P(l, s, t)$. In practice, however, this is numerically very demanding, so we confine ourselves to the study of three unconditional distributions as follows.

(1) $P(l)$ – the probability density of linear sizes irrespective of the avalanche's lifetime or energy release; that is, $P(l) = \int ds \int dt \, P(l, s, t)$.
(2) $P(s)$ – the probability density of energy releases irrespective of the avalanche's linear size or lifetime; that is, $P(s) = \int dl \int dt \, P(l, s, t)$.
(3) $P(t)$ – the probability density of lifetimes irrespective of the linear size or the energy release; that is, $P(t) = \int dl \int ds \, P(l, s, t)$.

The model is said to exhibit critical behavior because these three probability densities all follow power laws:

$$P(l) \sim l^{-\lambda},$$
$$P(s) \sim s^{-\tau}, \tag{4.16}$$
$$P(t) \sim t^{-\alpha}.$$

4.2.4 Numerical Results: Distribution Functions

Because the different properties of the avalanches (lifetime, linear extension, total number of topplings) are connected, we must consider a rather large number of conditional expectation values (Christensen et al. 1991). For example, consider the subset of avalanches whose lifetimes T are exactly equal to t. We may then ask how the average size S of this subset of avalanches depends on the value of T. If no scales characterize the model then we would expect the connection between S and T to be described by a scale-independent relationship, that is, a power law. To completely describe the model, we consider the following list of scaling exponents:

$$E[S \mid T = t] \sim t^{\gamma_1},$$
$$E[T \mid S = s] \sim s^{1/\gamma_1},$$

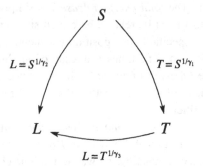

Figure 4.5. Transformations between the size, the lifetime, and
the linear extent of the avalanches.

$$E[S \mid L=l] \sim l^{\gamma_2},$$
$$E[L \mid S=s] \sim s^{1/\gamma_2},$$
$$E[T \mid L=l] \sim l^{\gamma_3},$$
$$E[L \mid T=t] \sim t^{1/\gamma_3},$$

(4.17)

where we have made use of the standard definition of conditional expectation values. For example,

$$E[S \mid T=t] = \sum_s sP(S=s \mid T=t).$$

(4.18)

In the definition of the scaling exponents in (4.17) we have made use of reciprocal relationships between some of the quantities. This is the same as assuming that the stochastic variables S, T, and L – representing the size, lifetime, and linear extent of the avalanches – can be connected by transformations: $T = S^{1/\gamma_1}$, $L = S^{1/\gamma_2}$, and $L = T^{1/\gamma_3}$, where we must have $\gamma_2 = \gamma_1\gamma_3$ (see Figure 4.5). Although this is not strictly true, the simulations of the model show that (say) the conditional density $P(S = s \mid T = t)$ has sufficiently narrow support to make it meaningful to talk about a characteristic size $S = s$ for all avalanches of duration $T = t$.

A set of scaling relations between the exponents defined in (4.16) and (4.17) are readily obtained from the following identity:

$$\int dy\, E[X \mid Y=y]P(Y=y) = \int dz\, E[X \mid Z=z]P(Z=z),$$

(4.19)

which is fulfilled for any set of three stochastic variables X, Y, Z. Let $X = S$, $Y = T$, and $Z = L$, and assume that $L = T^{1/\gamma_3}$. Then (4.19) produces the scaling relation

$$\alpha = 1 + \gamma_1 + \frac{\lambda - \gamma_2 - 1}{\gamma_3}. \tag{4.20}$$

Next, for $X = T$, $Y = S$, and $Z = L$ we obtain

$$\tau = 1 + \frac{1}{\gamma_1} + \frac{\lambda - \gamma_3 - 1}{\gamma_2}, \tag{4.21}$$

and finally from $X = L$, $Y = S$, and $Z = T$ we derive

$$\tau = 1 + \frac{1}{\gamma_2} + \frac{(\alpha - 2)\gamma_3 - 1}{\gamma_1 \gamma_3}. \tag{4.22}$$

Since $\gamma_2 = \gamma_1 \gamma_3$, (4.20) and (4.22) reduce to

$$\alpha = 1 + \frac{\lambda - 1}{\gamma_3}, \qquad \tau = 1 + \frac{\lambda - 1}{\gamma_2}. \tag{4.23}$$

These scaling relations are found to be consistent with numerical simulations of the distribution functions in two to five dimensions. The exponents depend on the dimension when $d < 4$. Within the range of numerical accuracy, the exponents for dimensions 4 and 5 are identical. This indicates that the critical dimension $d_c = 4$. The *critical dimension d_c* of a system is defined as the dimension above which the critical exponents become independent of the dimension and so assume the value calculated from mean field theory (see Section 5.2). The mean field value for the avalanche size exponent is found to be $\tau = 3/2$. This value is (within numerical accuracy) equal to the value obtained from simulations in four and five dimensions. The value of $\tau \simeq 1.3$ obtained from simulations in three dimensions (Grassberger and Manna 1990; Christensen et al. 1991) is significantly different from the mean field value of $3/2$. Hence we conclude that $d_c = 4$.

4.2.5 Power Spectrum

In BTW (1987) it was claimed that the power spectrum of the activity of the sandpile cellular automaton had the characteristic $1/f$ behavior. It was soon realized that this was, in fact, not quite right (Jensen et al. 1989; Kertész and Kiss 1990). In this section, some of the details involved in the calculation of the power spectrum are spelled out. We shall consider two different signals: first, the total flow down the slope of the pile when the pile is driven by adding extra grains at random positions of the surface of the pile. This signal turns out to have a $1/f^2$ spectrum in all dimensions. Secondly, we will consider the spectrum of the total amount of "z-value" in the heap when the pile is driven by pouring extra grains onto the pile only along the closed walls. This signal has a $1/f$ spectrum.

The easiest way to build up good statistics for making connections to the distribution of lifetimes is to assume that the activity of the pile, when driven by a continuous low influx of randomly positioned extra grains, is identical to a random linear superposition of avalanche signals. Indeed, simulations appear to justify this assumption (Jensen et al. 1989).

We therefore construct the signal of flow or dissipation in the following way. Choose a position at random, and add an extra grain of sand at this position. Imagine that the perturbation releases an avalanche. We then monitor the dissipation rate† $f_A(\tau)$ at each subsequent time step τ after initialization of the specific type-A avalanche, where A characterizes the avalanche in terms of (say) lifetime, total number of topplings, or linear extent. After building up a library of temporal avalanche profiles $f_A(\tau)$, we construct the total dissipation rate as a linear superposition of the time signals $f_A(\tau)$ started at random times:

$$J(\tau) = \sum_{\tau_i < \tau} f_{A_i}(\tau - \tau_i). \qquad (4.24)$$

Here τ_i is a set of random uniformly distributed starting times, and A_i denotes the type of the avalanche started at time τ_i. By construction, there are no correlations between the individual avalanche signals contributing to the sum in (4.24). The power spectrum of the signal $J(\tau)$ becomes the weighted sum of the spectra of the individual avalanches:

$$S_J(f) = \nu \sum_A P(A)|\hat{f}_A(f)|^2$$
$$= \frac{\nu}{\pi^2 f^2} \int_0^\infty d\tau \, \Lambda(\tau) \sin^2(\pi f \tau), \qquad (4.25)$$

where we have introduced the weighted lifetime distribution

$$\Lambda(t) = \frac{1}{t^2} \int_0^\infty ds \, P(s,t)s^2 \qquad (4.26)$$

(see Appendix D for details). The frequency dependence of $S_J(f)$ is determined by the distribution of weighted lifetimes. A behavior of the form

$$\Lambda(t) \sim t^a \exp(-t/t_0) \quad \text{for } t > t_0 \qquad (4.27)$$

leads to a power spectrum with the following algebraic frequency dependence:

$$S_J(f) \sim \begin{cases} f^{-(3+a)} & \text{if } a < -1, \\ f^{-2} & \text{if } a > -1. \end{cases} \qquad (4.28)$$

† We define the *dissipation rate* $f_A(\tau)$ as the number of overcritical sites being relaxed at time step τ.

In all dimensions simulated, the exponent a is found to be larger than -1 and the power spectrum of the flow down the slope is accordingly of the form $S_J(f) \sim f^{-2}$.

One may ask why this should be less interesting than a spectrum $S_J(f) \sim f^{-\beta}$ with an exponent $\beta \simeq 1$. In itself, a genuine exponent $\beta = 2$ is certainly interesting. However, the observed spectra, which the SOC approach claimed to describe, have exponents β in an interval of about $1/2 \leq \beta \leq 3/2$, well below the value of 2. Of course there is an abundance of $1/f^2$ spectra observed in nature. The point, though, is that in most cases these spectra are expected to be the high-frequency limit of some Poisson process with a spectrum $1/(1 + (f/f_c)^2)$. If this is in fact the case then no correlations exist in the long-time–low-frequency limit, $t > 1/2\pi f_c$. This is the reason for the prejudice that considers $1/f^2$ spectra to be less exciting than $1/f$ spectra. But this is only a fair judgment when the $1/f^\beta$ behavior with $\beta \simeq 1$ persists down to very small frequency values. Convergence entails that $1/f^\beta$ behavior must eventually be cut off for frequencies below a certain frequency f_{cut}. The interesting point is that this frequency often corresponds to time scales many orders of magnitude longer than the time scale of the individual microscopic processes. A spectacular example is the resistivity fluctuation in conductors, where $1/f$ behavior has been observed over more than six orders of magnitude (Weissman 1988).†

After all these remarks, we are now able to judge if the $1/f^2$ spectrum for the sand-flow signal in (4.24) is an example of an interesting $1/f^2$ signal. The answer is that this signal is just as interesting as the algebraic distribution of avalanche lifetimes. The $1/f^2$ behavior will be cut off below a frequency $f_{\text{cut}} \sim 1/T_{\text{max}}(L)$, where $T_{\text{max}}(L)$ is a measure of the longest avalanche lifetime possible for the given system size L. Hence the $1/f^2$ behavior can be made to span as many decades of frequencies as desired simply by increasing the size of the system. Thus the $1/f^2$ spectrum in the BTW model is not merely the high-frequency tail of the spectrum of the Poisson process. We know this with certainty because we know how the signal $J(\tau)$ was created. In this sense the observed $1/f^2$ spectrum is as interesting as any other power law behavior. However, it is obvious that the $1/f^2$ behavior cannot be claimed to explain why we so often observe $1/f$ spectra.

It is curious that the BTW sandpile does after all contain a $1/f$ power spectrum if driven in a specific way. Even more remarkable is that the $1/f$ regime is indeed a low-frequency property – one that is characteristic of frequencies *smaller* than the time scale $T_{\text{max}}(L)$ corresponding to the longest avalanche

---

† A note of caution is appropriate. Spectra with $\beta > 1$ might not correspond to stationary processes, in which case extra care must be exercised when dealing with the power spectrum. See e.g. Bonabeau and Lederer (1994).

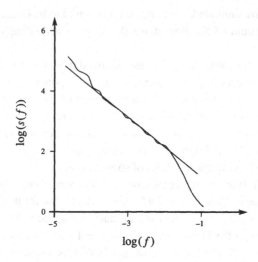

Figure 4.6. The power spectrum of the total z-value of the edge-driven sandpile. (Sketch of data presented in Jensen 1991.) The straight line has a slope equal to -1. Deviation from $1/f$ behavior at high frequencies occurs at a frequency given by the maximum lifetime of an avalanche for the considered system size.

lifetime! This spectrum is generated in the following way. Let $N(t)$ denote the sum over the lattice of the dynamical variable at a given time step; that is, $N(t) = \sum_i z_i(t)$. We imagine that the model describes a pile of sand created in the corner of a square table (see Figure 4.2). The table has walls along two of the edges, say $x = 0$ and $y = 0$. Along these two edges we apply the closed boundary condition. Along the edges $x = L$ and $y = L$ we use the open boundary condition. The system is driven by adding sand at random positions chosen along the closed boundary. Notice that we refrain from adding any sand to the interior of the surface of the pile. The power spectrum of the signal $N(t)$, which we might think of as the total mass on the table at a given time, has the form shown in Figure 4.6. For frequencies $f < 1/T_{max}(L)$ the spectrum behaves as $S(f) \sim 1/f$.

This type of $1/f$ behavior can be obtained from a boundary-driven diffusion equation without any bulk noise. For details, see Section 5.2.3.

4.3 Earthquake Models: Nonconservative Models

The cellular automata model described in the previous section was inspired by the dynamics of sandpiles. Because the number of sand grains are conserved as they flow down the slope of a sand heap, the numerical algorithm considered was designed to conserve the dynamical variable during each update (except at the boundary of the system). If the dynamical variable is imagined to represent quantities other than sand grains then there is no reason why the update should necessarily conserve the dynamical field. In their 1992 paper, Olami, Feder, and Christensen (OFC) had a model of sliding tectonic sheets in mind when they viewed the dynamical variable as the local force or stress. In this case, conservation may be broken. Since the justification for the model is largely heuristic anyway, for simplicity we will first define the algorithm and then later discuss how the model is related to physical systems.

Consider an open d-dimensional cubic lattice of size $N = L^d$. To each lattice point i we ascribe a *real* dynamical variable E_i. We think of this field as representing an energy or a force, and to be specific we will talk about the *energy* E_i of site i. All sites are driven at the same rate. That is, during every time step,

$$E_i \rightarrow E_i + \nu \quad \text{for } i = 1, \ldots, N. \tag{4.29}$$

This mode of driving is used only as long as $E_i < E_c$ for all sites, where E_c is some threshold. When a site becomes unstable, $E_i \geq E_c$, the homogeneous drive is switched off and the sites are relaxed according to the following local relaxation rule:

$$E_i \geq E_c \Rightarrow \begin{cases} E_i \rightarrow 0, \\ E_{nn} \rightarrow E_{nn} + \alpha E_i, \end{cases} \tag{4.30}$$

where nn denotes the neighbor sites of the overcritical site number i. During each iteration of (4.30), an amount of energy is lost from the system – namely, the difference between the energy that disappears at the center site E_i and the energy αE_i added to each of the q_i neighbor sites. Here q_i denotes the coordination number of the ith site in the lattice: $q_i = q_c = 2d$ for a bulk site in a d-dimensional cubic lattice with nearest neighbor assignment. The coordination number for sites at the surface of the system is smaller and varies in an obvious way for surface, edge, and corner sites. The difference between lost and added energy is accordingly given by

$$E_{\text{dis}} = (q_i \alpha - 1) E_i. \tag{4.31}$$

Only if one sets α equal to $1/q_c$ will the algorithm conserve the dynamical variable in the bulk. The update in (4.30) is iterated until all site variables E_i are yet again undercritical.

Let us point out two important differences between the OFC model and the BTW model of Section 4.2. First, the variable E_i is a real variable; hence the model can be driven infinitely gently in the sense that we can let $\nu \rightarrow 0$ in (4.29). This limit is most efficiently obtained in the following way. When an avalanche (i.e. a chain of iterations of (4.30)) has come to an end, one locates the site with the largest E-value; call it E^*. Add now to all sites the difference $E_c - E^*$. This will make the site with E^* critical, and one can then start iterating (4.30). More than one site can assume the maximum value E^*, in which case there is an ambiguity concerning which site to update first. However, a tiny random amount added to each site in between the avalanches will resolve this ambiguity and apparently has no influence on the results (Middleton and Tang 1995). Alternatively, one could pick a site at random among the set of all sites assuming the value E^*.

The second important difference is that the update in the BTW model (see (4.9)) is strictly conservative. Conservation was expected to be an important prerequisite for models exhibiting power law behavior (see Grinstein, Lee, and Sachdev 1990). The critical behavior of the OFC model in the nonconservative regime $\alpha < 1/q_c$ came as a surprise.

4.3.1 Criticality of the OFC Model

Let us now discuss the numerical evidence for critical behavior in the OFC model. The neighbor assignment in the update algorithm in (4.30) can be interpreted in different ways. A natural assignment of neighbors is obviously the standard *nearest neighbor* interpretation. Another assignment, frequently used in statistical mechanics, is a *random* neighbor assignment. One can once and for all choose at random a number of sites and assign these sites as "neighbors" to site i. Alternatively, one can assign a new list of neighbors every time one performs the update in (4.30); in this case one talks about *annealed* random neighbors. We shall concentrate on the nearest neighbor version of the model and the random neighbor version with annealed random neighbors. The virtue of the latter is that the reassignment of neighbors in every time step destroys spatial correlations. A comparison of the random neighbor version and the nearest neighbor version of the model can therefore help isolate the effect of spatial correlations from other effects.

4.3.2 Nearest Neighbor OFC Model

We start from, say, a configuration consisting of randomly assigned E_i values with $E_i \in [0, E_c)$. The algorithm is then iterated until a stationary state is reached – that is, a state in which the distribution of avalanche sizes or avalanche lifetimes does not change with time. The stationary state is reached very slowly. About 10^8 or 10^9 avalanches are needed to reach the stationary state for two-dimensional lattices of size $L \sim 10^2$. This number increases as α is decreased from $1/4$ (see Grassberger 1994). This is a serious problem, which

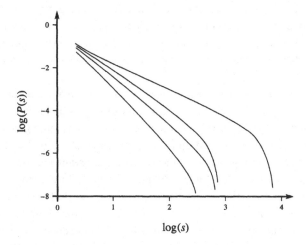

Figure 4.7. Probability density of avalanche sizes in the OFC model for
different values of the conservation level. (Sketch of data
presented in Olami, Feder, and Christensen 1992.)

makes it difficult to assess the properties of the model for small α values. From
the probability density of avalanche sizes, one clearly gets the impression that
the model remains critical for α values smaller than $1/4$ but not too close to 0
(see Figure 4.7).

It was originally concluded by OFC (see Christensen and Olami 1992a) that
a transition from critical behavior for $\alpha > \alpha_c$ to noncritical behavior for $\alpha <
\alpha_c$ took place at a value $\alpha_c \simeq 0.05$. They reached this conclusion by looking
at log-log plots of the avalanche size distribution and observing that no part of
the graphs could be fitted to a straight line when α is smaller than about 0.05.
A similar conclusion was also reached by Grassberger (1994) and by Corral
et al. (1995), although these authors instead estimate α_c to be about 0.16 to
0.18. The nature of the supposed change in behavior at α_c is not clear. Grassberger (1994) expected the change to be gradual, yet Corral et al. (1995) presented arguments in favor of a sharp transition. Middleton and Tang (1995)
argued that, in reality, no transition occurs for finite α_c; that is, $\alpha_c = 0$. The
model is definitely noncritical for $\alpha = 0$, since all avalanches are restricted
to a single site and no interaction is possible between sites when $\alpha = 0$; see
(4.30).

It is difficult to determine the value of α_c from numerical simulations because of the very long transient period before the stationary state is reached.
The critical state migrates in from the open boundary (Middleton and Tang

1995) at a speed that decreases with α. The critical behavior is entirely lost if, instead of open boundaries, one uses a periodic boundary condition. No single time scale determines the dynamics of the system with open boundaries (for $\alpha > \alpha_c$), whereas the sites of the periodic system tend to synchronize into a periodic state.

The tendency to synchronize is well known for coupled oscillators. Even a minute coupling can lead to perfect synchronization, as described in the seventeenth century by Huygens. He noted that two clocks hanging next to each other on his wall always ended up moving in phase, and found that this was due to the slight coupling caused by the vibrations transmitted through the wall. The site variables E_i of the OFC model are like coupled oscillators that oscillate between 0 and E_c (though overcritical values $E_i > E_c$ do occur during the evolution of the avalanches). The propensity toward synchronization is partly responsible for the critical behavior observed in the nonconservative regime $\alpha < 1/q_c$. The synchronization builds up spatial correlations, and the spatially homogeneous system with periodic boundary conditions synchronizes into a periodic state (Socolar, Grinstein, and Jayaprakash 1993; Middleton and Tang 1995). The inhomogeneity introduced by an open boundary marginally breaks the synchronization. The phase locking is disturbed enough to destroy the periodic behavior, but the disturbance does not destroy the correlations, which are left sufficiently intact to allow the critical state to exist.

Marginal synchronization is not the only mechanism responsible for the existence of criticality in the nonconservative regime. As discussed in the next section, this can be seen from studies of the random neighbor version of the OFC model.

4.3.3 Random Neighbor OFC Model

The random neighbor assignment completely prevents synchronization. Nevertheless, a regime of power law behavior can be identified. The random neighbor OFC model is defined like the nearest neighbor model discussed in Section 4.3.2, except that neighbors are assigned in a different way. Now, when we update a site i, we choose (at random) q_i different sites among the $N - 1$ remaining sites. By choosing q_i appropriately we can take the effect of the open boundary into account. This is done in the following way. When a site is to be relaxed according to (4.30) we consider the site to be a "bulk" site (i.e., $q_c = 2d$) with probability $p_{\text{bulk}} = (L - 1)^d/L^d$ or a "surface" site ($q_i = 2d - 1$) with probability $1 - p_{\text{bulk}}$. Because a new assignment of neighbors is made in each update, the spatial lattice structure is lost. Hence spatial correlations cannot exist, and synchronization cannot occur.

The size of an avalanche, measured as the number of sites updated during the avalanche, can be measured in exactly the same way as in the nearest neighbor model. Hence, we can determine the probability density of avalanche sizes $P_\alpha(s)$ for a given value of α. From finite size scaling of data sets measured in two-dimensional simulations it was shown that, for $\alpha > \alpha_c \simeq 2/9$, the density follows a power law: $P_\alpha(s) \sim 1/s^\tau$, where $\tau \simeq 3/2$ (Lise and Jensen 1996). The power law behavior is cut off only by the finite size of the system for $\alpha > \alpha_c$. For $\alpha < \alpha_c$, the size distribution is well approximated by

$$P_\alpha(s) \sim s^{-3/2} \exp(-s/s_0). \tag{4.32}$$

Here the characteristic size s_0 diverges according to $s_0 \sim (\alpha_c - \alpha)^{-\nu}$, where $\nu \simeq 3/2$ when α approaches the critical value α_c. Since s_0 is independent of L, we must consider the system as noncritical for $\alpha < \alpha_c$. The critical behavior in the nonconservative regime $[\alpha_c, 1/4)$ is not caused by any kind of marginal synchronization. Rather, the critical behavior can be understood as a kind of dynamical conservation during the evolution of the avalanches. We shall return to this interpretation in connection with the mean field discussion of this model; see Section 5.2.2. (See also the note on p. 124.)

The transition from critical to noncritical behavior at α_c in the random neighbor model seems to be very well established numerically. The reason that the numerical results are relatively unproblematic in this version of the model is that α_c is so high. In the nearest neighbor model, the transition (if any) occurs at a much lower α value where it is difficult to obtain high-quality statistics, say for $P_\alpha(s)$, over a significant range of avalanche sizes. This is because, in the nearest neighbor version of the model, $P_\alpha(s) \sim s^{-\tau_\alpha}$ where τ_α increases with decreasing α. The next section discusses the size distribution of, as well as temporal fluctuations in, the nearest neighbor model.

4.3.4 Distributions and Fluctuations in the Nearest Neighbor OFC Model

The exponent τ_α for the algebraic behavior of the avalanche distribution $P_\alpha(s)$ in the nearest neighbor OFC model exhibits a dependence on the level of conservation α. Christensen and Olami (1992a) were able to analyze their simulations in terms of the finite size scaling form

$$P_\alpha(s, L) = L^{-\beta} g_\alpha(s/L^\nu), \tag{4.33}$$

where g_α is an α-dependent scaling function common to all system sizes. The size of the considered system enters only through the explicit L-dependence in (4.33). If one finds that data from simulations of different system sizes can all be fitted by (4.33) for one specific choice of exponents β and ν, it is of course tempting to believe that one has full control of the size dependence. Furthermore, a consistency check can be obtained when the measured probability densities behave like power laws $P_\alpha(s) \sim s^{-\tau_\alpha}$. In this case $\tau_\alpha = \beta/\nu$, which can be seen as follows. Write (4.33) as a power law multiplied by a scaling function that takes care of the size dependence. This yields

$$P_\alpha(s, L) = L^{-\beta} g_\alpha(s/L^\nu)$$

$$= L^{-\beta} \left(\frac{s}{L^\nu}\right)^{-\beta/\nu} \left(\frac{s}{L^\nu}\right)^{\beta/\nu} g_\alpha\left(\frac{s}{L^\nu}\right)$$

$$= s^{-\beta/\nu} \tilde{g}_\alpha\left(\frac{s}{L^\nu}\right), \tag{4.34}$$

where we have introduced a new scaling function $\tilde{g}_\alpha(x) = x^{\beta/\nu} g_\alpha(x)$.

For the somewhat small system sizes ($25 \leq L \leq 45$) considered by Christensen and Olami, the scaling form in (4.33), the power law behavior, and the consistency check all work nicely for $P_\alpha(s, L)$ when α is not too far from $1/4$. Unfortunately, the behavior of the OFC model turns out to be more complicated when larger system sizes are simulated (Grassberger 1994). This illustrates a concern that must always be kept in mind when doing computer simulations. Namely, is the observed behavior within the scaling regime? That is, will the behavior of larger systems (than are possible to consider by available computer resources) be connected to the simulated behavior by a simple "blow-up" factor? Or does a crossover to a different type of behavior occur at some size of the system?

That τ_α does depend on α is clear (Middleton and Tang 1995), but the simple finite size dependence described by (4.33) does not survive for system sizes $L > 100$. The model is complicated because the behavior in the bulk is very different from the behavior closer to the edge (see Section 4.3.2). For small systems, the edge behavior dominates. As system size increases, the bulk becomes more important. The inhomogeneous behavior of the system makes it impossible to connect one system size to another by simple scaling, since the bulk behavior scales differently from the edge behavior. The lack of a well-understood size dependence of $P_\alpha(s, L)$ means that the detailed behavior of the exponent τ_α as a function of α is still unknown.

Until now we have discussed only the spatial properties of the individual avalanches, namely, the distribution of the number of sites involved in an avalanche. Of course, one can also measure the duration of the avalanches – that is, the number of updates needed to make all sites of the system undercritical after an avalanche is released at some site i with $E_i > E_c$. The probability density of the avalanche lifetime is, as expected, also a power law. Another interesting property of the model is the temporal organization of avalanches. Recall the discussion of earthquakes in Section 3.6. It is equally important to understand the Gutenberg–Richter law for avalanche sizes as it is to understand the Omori law for the distribution of aftershocks. One can study the interval between avalanches in the OFC model. The dynamics of the model involves two time scales: one for the external driving defined in (4.29); another for the internal relaxation during the evolution of an induced avalanche, given by (4.30). As usual, for the models considered in connection with SOC, we think of systems for which there is a separation between the time scale of the external drive and the time scale of the internal response. Consider again earthquakes. The rate of motion of the tectonic plates is centimeters per year. On the other hand, the release of stress during earthquakes occurs within a few seconds or minutes. With all this in mind, it is natural to consider the relaxation via iteration of (4.30) as instantaneous on the scale of the driving, (4.29) (see Olami and Christensen 1992). The constant rate of driving can be expressed as

$$\frac{dE_i}{dt} = 1, \tag{4.35}$$

using an appropriate definition of the time unit. Equation (4.35) allows us to define the time span between successive releases of avalanches as

$$\Delta = E_c - \max_i E_i, \tag{4.36}$$

where $\max_i E_i$ denotes the largest E-value on the lattice immediately after the relaxation (by (4.30)) of the previous avalanche has come to an end. Following this procedure, we can construct a temporal record $S(t)$ of the "avalanche dynamics" in the model. We define $S(t)$ in the following way. The signal is always zero except at those instances when avalanches occur, in which case $S(t)$ is equal to the size s of the avalanche released at t. That is (as depicted in Figure 4.8), the signal $S(t)$ consists of a set of delta spikes of height equal to the size of the avalanches and of separations given by (4.36). The time signal $S(t)$ exhibits interesting fluctuations.

One can study the probability density $P_{s_0}(\delta)$ of intervals δ between avalanches larger than a given size s_0. Assume for a moment that the avalanches greater than s_0 occur randomly in time at an average frequency $\nu = 1/\langle \delta \rangle$, as in a Poisson process. To be precise, this means that the probability that an avalanche is released during the small time interval dt is equal to $\nu \, dt$. In this case, the probability density of the waiting times between avalanches is exponential:

$$P_{s_0}(\delta) = \frac{1}{\langle \delta \rangle} \exp\left(\frac{-\delta}{\langle \delta \rangle}\right). \tag{4.37}$$

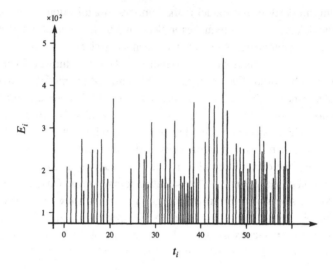

Figure 4.8. Energy released by avalanches as a function of time.
(Sketch of data presented in Olami and Christensen 1992.)

This is a standard result, and we have discussed the derivation in detail in Appendix D.

The functional form of $P_{s_0}(\delta)$ found in simulations of the OFC model depends on the size of s_0. For small threshold sizes, the probability density is in fact found to be exponential. However, when we focus on the temporal separation of large avalanches by choosing s_0 large, the functional form of $P_{s_0}(\delta)$ changes to a power law (see Christensen and Olami 1992b). This power law distribution is caused by a clustering of the large avalanches. Namely, the excess number of (large) avalanches following a large avalanche is found to decay like $1/t$, where t is the time elapsed since the occurrence of a certain large avalanche. This follows from the simulations of the behavior of the quantity

$$g_{s_0}(t) = \langle n(t) \rangle_{t_0} - t\bar{n}, \qquad (4.38)$$

where $\langle n(t) \rangle_{t_0}$ denotes the number of avalanches (larger than s_0) during the interval $[t_0, t_0 + t]$ averaged over t_0. The average number of avalanches (larger than s_0) per time is denoted by $\bar{n}$ (see Olami and Christensen 1992). The time derivative of $g_{s_0}(t)$ measures the deviation of the frequency of avalanches from the average frequency. For large s_0, one finds approximately $g_{s_0}(t) \sim \ln(t)$ for values of t that are not too large. Hence, the excess number dg_{s_0}/dt of avalanches occurring after a large avalanche decays like $1/t$.

Because they are correlated in space, it seems reasonable that large avalanches in the OFC model are also correlated in time; two large consecutive avalanches are more likely to be correlated since they overlap in space. In the same way it is not too surprising that the small avalanches lack temporal correlations, since they may easily occur in regions that are well separated in space.

Exercises. (1) Assume that avalanches larger than s_0 occur at random and in an uncorrelated fashion. Calculate $g_{s_0}(t)$.

(2) Study the fractal dimension (by box counting) of the support of the time signal $S_{s_0}(t)$ constructed from $S(t)$ by including only avalanches larger than s_0. See Olami and Christensen (1992).

(3) Study the time signal $S(t)$ in the random neighbor version of the OFC model. See Section 4.3.3.

4.3.5 The Effect of Disorder on the OFC Model

There are many ways of introducing disorder into the OFC model. The first question is, of course, Why bother? The answer is that real physical systems are never completely homogeneous. Spatial and temporal variations in the properties of systems always occur. Depending on the kind of system to which the OFC model will be applied, one can think of a variety of ways of introducing spatial and/or temporal variations in the parameters defining the model. The effect of different types of disorder has been studied numerically. In general, the results have not been very conclusive, and more work is certainly needed. Difficulty arises when one tries to understand the effect of disorder for different levels of conservation, that is, for different values of the parameter α. Most numerical studies use double logarithmic plots of various distribution functions to probe critical behavior. By this method it is difficult even for the pure model to determine how far down in α-values the critical behavior persists (see Section 4.3.2). For a specific choice of parameters, the criticality of the model is numerically most often studied by probing the scaling of distribution functions with increasing system size. The results, however, are seldom unambiguous. Consider, for example, the probability density of avalanche sizes. Assume that we obtain a functional form such as

$$P(s) \sim s^{-\tau} \exp(-s/s_0). \tag{4.39}$$

Assume that s_0 increases with increasing system size as, say, some power $s_0 \sim L^{\nu}$ of the linear extension of the system. In this case, one will probably feel rather safe in concluding that the system exhibits true critical behavior without any intrinsic characteristic scale (except for the scale set by the size of the system). Now one can be fooled, of course. It could be that s_0 increases with L for (say) $L < L^*$ but then becomes independent of L for $L > L^*$. There is really no guarantee against this type of unpleasant behavior. If L^* is large compared to what can be economically simulated with available resources, one may be led mistakenly to conclude that the system is in a critical state even though this is not actually the case when an intrinsic scale (e.g., L^*) does exist.

Unfortunately, this problem has in the past led to erroneous conclusions concerning the behavior of various models considered as examples of SOC, and it is likely to do so in the future as well. In the field of equilibrium statistical mechanics, a similar problem arises when one wants to determine the order of a phase transition. For continuous transitions, the correlation length $\xi(T)$ diverges as the temperature T is approaching the critical temperature T_c. For a discontinuous, or first-order, transition, the correlation length increases as T is tuned toward T_c, but remains finite at $T = T_c$. If $\xi(T_c)$ is much larger than the system size of the simulations then – from the finite size scaling of the simulated data – one might easily draw the fallacious conclusion that the transition involves a diverging length scale and is therefore continuous. A way around this problem was found by Lee and Kosterlitz (1991), who showed how the nature of the phase transition could be obtained from the probability distribution of the system's total energy. The method turns out to work even if the considered system size is significantly smaller than $\xi(T_c)$. A similar method is not currently known for SOC systems. These difficulties must be kept in mind during the following discussion.

Let us consider the effect on the critical behavior of the OFC model when different types of disorder are introduced. First of all, the drive can vary in space. For instance, instead of (4.29) one could drive the model in the same way as the BTW model is driven (Christensen 1992; Lise and Jensen 1995). Choose at random a site i and perform the update

$$E_i \rightarrow E_i + \delta E. \qquad (4.40)$$

Here δE can be a fixed value or one could, in every update, choose δE at random from some distribution $P(\delta E)$. Simulations show that a cutoff is introduced in the distribution of avalanche sizes $P(s)$ for large δE. The cutoff is independent of the system size and thus destroys the algebraic behavior that is the hallmark of criticality. This result is unproblematic. One can easily simulate different system sizes and observe that $P(s)$ cuts off at a value s_0 that is independent of the system size when δE is sufficiently large. In other words, there is no increase with system size in the interval over which $P(s)$ exhibits power law behavior.

As δE becomes smaller, $P(s)$ begins to depend on the system size L, at least for small L. It is difficult to decide whether the cutoff $s_0(\delta E) \rightarrow \infty$ for $\delta E \rightarrow$

δE_{thr} for some nonzero value of δE_{thr}. Optimally one would like to be able to identify a scale s_0 that diverges as

$$s_0 \sim (\delta E - \delta E_{\text{thr}})^{-\nu}. \tag{4.41}$$

This kind of behavior is seen in the random neighbor version of the OFC model (see Section 4.3.3) when the critical conservation level $\alpha_c \simeq 2/9$ is approached from below. Unfortunately, a similarly clear scaling of s_0 has not yet been identified for any value of δE_{thr}. It may be that random drive, of the form in (4.40), destroys the critical behavior no matter how small δE is. This is what renormalization group calculations indicate (see Section 5.4).

The effect of randomizing the threshold E_c in (4.30) has been studied in two dimensions by Jánosi and Kertész (1993). Each site i is ascribed a threshold value E_i^c drawn from a uniform distribution on $[E_c - \delta E_c, E_c + \delta E_c]$. For $\alpha < 1/4$ they concluded that any nonzero value of δE_c induces a length scale and that the size distribution becomes exponential. The conservative case $\alpha = 1/4$ is insensitive to the randomness. Unfortunately, Jánosi and Kertész did not investigate the divergence of the characteristic length as $\delta E_c \to 0$.

Let us finally discuss the effect of introducing a random element into the redistribution algorithm (4.30). We replace (4.30) by

$$E_i \geq E_c \Rightarrow \begin{cases} E_i \to 0, \\ E_{nn} \to E_{nn} + \alpha_i E_i, \end{cases} \tag{4.42}$$

where each site has its own redistribution parameter $\alpha_i = \alpha + \delta_i$. The deviation δ_i is a random number uniformly drawn from the interval $[-\delta, \delta]$. To ensure that no site generates energy, α_i is restricted to the interval $0 < \alpha_i < 0.25$ (in two dimensions). This is obtained by discarding δ_i if $\alpha + \delta_i > 0.25$. The values α_i are fixed during the simulation. This is called *quenched disorder.*

The SOC behavior of the pure model is observed to survive if only a small concentration of sites are assigned an α_i value different from α (Ceva 1995). The case where all sites are assigned a random value α_i has been studied by Mousseau (1996), who concluded from simulations of the avalanche size distribution that four different regions exist in the parameter space (α, δ). As sketched in Figure 4.9, SOC behavior is claimed to survive in a region close to the α-axis when α is larger than about 0.1. Above this region (i.e., for larger δ-values), the model tends to synchronize and the broad algebraic distribution of avalanches is lost. Increasing the amplitude of the randomness δ even more brings the system back into a region (number III in Figure 4.9) where the distribution of avalanche sizes $P(s)$ becomes broad, with a power law regime for $0 < s < s_0$. It is not clear from the simulation data that the cutoff s_0 of the power law scales to infinity as $L \to \infty$ in region III. This is in contrast to the situation in region

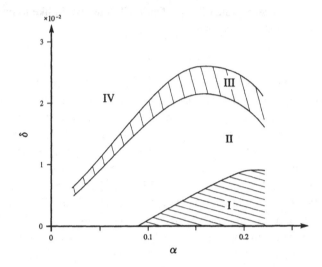

Figure 4.9. Proposed conservation–disorder phase diagram of the OFC model.
(Sketch of data from Mousseau 1996.)

I, where s_0 seems to increase with the system size. Finally, one leaves region III as δ is increased further and so enters region IV, a region of exponential distributions. Unfortunately, no attempt was made to underpin these suggestions by serious finite size scaling analysis.

We have shown in this section that the change in behavior of the OFC model as disorder is introduced is rather subtle. Definite conclusions via numerical simulations are elusive because it is difficult to ensure that the considered system sizes are within the scaling regime. For the SOC behavior of the model to be of relevance to real physical systems, it is necessary that the model exhibit a certain amount of robustness in the face of disorder. We cannot yet say if such robustness actually exists.

Exercises. Use the generalized mean field theory of the branching ratio (described in Section 5.2.2) to discuss the effect of disorder.
(a) Consider disorder (or fluctuations) in the drive; see (4.40).
(b) Consider disorder in the thresholds.
(c) Consider disorder in the distribution parameter; see (4.42).

4.3.6 Physical Relevance of the OFC Model

The modeling of earthquakes was part of the original motivation for the form of the dynamics given by (4.29) and (4.30) (Olami et al. 1992); OFC sought

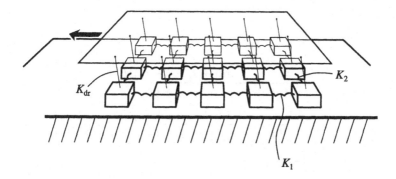

Figure 4.10. The Burridge–Knopoff spring–block model.

a simple representation of the dynamics of the so-called Burridge–Knopoff spring–block model (Burridge and Knopoff 1967). This is a mechanical model consisting of a two-dimensional square array of blocks (see Figure 4.10).

Each block is connected by elastic springs to its four neighbors. The blocks sit on a fixed plate and are coupled by leaf springs to a moving plate on top. The total force on the block at position (i, j) is given by

$$\mathbf{F}_{(i,j)} = \mathbf{f}^{dr}_{(i,j)} + \mathbf{f}_{(i+1,j)} + \mathbf{f}_{(i-1,j)} + \mathbf{f}_{(i,j+1)} + \mathbf{f}_{(i,j-1)}. \qquad (4.43)$$

The force $\mathbf{f}^{dr}_{(i,j)}$ is the driving force exerted by the leaf spring, and the forces $\mathbf{f}_{(i\pm1,j\pm1)}$ are those produced by the springs connecting the block at (i, j) to its nearest neighbors. The friction force exerted by the static plate is incorporated as the threshold criterion. If the force $\mathbf{F}_{(i,j)}$ is smaller than the static friction force then the block remains static. The block jumps ahead when $\mathbf{F}_{(i,j)}$ becomes larger than the static friction force. The block is supposed to jump to a new position of force balance given by $\mathbf{F}_{(i,j)} = 0$.

We need a bit of notation in order to write down the expression for the forces in terms of the displacements of the blocks (Leung, Müller, and Andersen 1997). Let K_1 and K_2 denote the spring constants of the springs along the x-direction and the y-direction, respectively. Similarly, let the equilibrium length of these springs be l_1 and l_2. The system may be prepared in a state of internal strain by assuming that the average distance between the blocks is $a_k \neq l_k$, $k = 1, 2$. Imagine for example that the whole array has been expanded a certain degree, $a_k > l_k$. Let $(x_{i,j}, y_{i,j})$ denote the displacement of the block (i, j) away from the strained equilibrium configuration, and assume that the springs follow Hook's linear force law. Then the force on block (i, j) is given by the following (rather awe-inspiring) expression:

$$F_{i,j}^x = f_{(i,j)}^{\mathrm{dr},x}$$

$$- K_1 \bigg\{ 2x_{i,j} - x_{i+1,j} - x_{i-1,j}$$

$$+ \frac{(a_1 + x_{i+1,j} - x_{i,j})l_1}{[(a_1 + x_{i+1,j} - x_{i,j})^2 + (y_{i+1,j} - y_{i,j})^2]^{1/2}}$$

$$- \frac{(a_1 + x_{i-1,j} + x_{i,j})l_1}{[(a_1 - x_{i-1,j} + x_{i,j})^2 + (y_{i,j} - y_{i-1,j})^2]^{1/2}} \bigg\}$$

$$- K_2 \bigg\{ 2x_{i,j} - x_{i,j+1} - x_{i,j-1}$$

$$+ \frac{(x_{i,j+1} - x_{i,j})l_2}{[(a_2 + y_{i,j+1} - y_{i,j})^2 + (x_{i,j+1} - x_{i,j})^2]^{1/2}}$$

$$+ \frac{(x_{i,j-1} - x_{i,j})l_2}{[(a_2 - y_{i,j-1} + y_{i,j})^2 + (x_{i,j} - x_{i,j-1})^2]^{1/2}} \bigg\}. \quad (4.44)$$

The y-component of the force is obtained from this expression by symmetry. This equation can be linearized in the displacements. Define $\mathbf{f}_{(i,j)}^{\mathrm{dr}} = (-K_{\mathrm{dr}}x_{i,j}, 0)$. The rearrangement needed in order to perform the slip of the block (i, j) defined by $\mathbf{F}_{i,j} \rightarrow 0$ is, in the linear limit, given by

$$F_{i\pm1,j}^x \rightarrow F_{i\pm1,j}^x + \alpha_1 F_{i,j}^x,$$

$$F_{i,j\pm1}^x \rightarrow F_{i,j\pm1}^x + S_2\sigma\alpha_1 F_{i,j}^x,$$

$$F_{i\pm1,j}^y \rightarrow F_{i\pm1,j}^y + \frac{S_1\alpha_2}{\sigma} F_{i,j}^y, \qquad (4.45)$$

$$F_{i,j\pm1}^y \rightarrow F_{i\pm1,j}^x + \alpha_2 F_{i,j}^y.$$

The internal strain is measured by $S_k = (a_k - l_k)/a_k$ for $k = 1, 2$. The anisotropy of the system is measured by $\sigma = K_2/K_1$ and $\kappa = K_{\mathrm{dr}}/K_1$. The redistribution parameters α_k are defined as

$$\alpha_1 = \frac{1}{2(1 + S_2\sigma) + \kappa} \quad \text{and} \quad \alpha_2 = \frac{1}{2(1 + S_1\sigma)}. \qquad (4.46)$$

This tour de force makes it transparent in what limit the OFC model corresponds to the Burridge–Knopoff model. The update algorithm defined in (4.30) is recovered in the limit $F_{i,j}^y \equiv 0$ and $S_2\sigma = 1$. The first condition is obtained (to first order) if the displacements in the y-direction can be ignored. The second condition, $S_2\sigma = 1$, corresponds to a rather ad hoc balance between the strain along the y-direction and the elastic anisotropy of the system. It is important to notice that the vector nature of (4.45) turns out to be unimportant as long as the driving is along one of the symmetry axes. Nor is the inclusion of nonlinear terms in the force important. However, the careful derivation of the force in (4.44) by Leung and collaborators has made it clear that the OFC model is related to a system with internal strain. Different α-values in the OFC model must be interpreted as representing differences in anisotropy as well as differences in the internal strain.

The physical meaning of the choice of boundary conditions imposed on the model is also interesting. Three different types of boundary conditions have been considered for the OFC model.

(1) *Free Boundary Conditions.* The blocks at the edge are considered to have three neighbors, while the blocks at the corners are considered to have two neighbors. The internal strain of the system is thought of as being sustained in some unspecified way – for instance, by appropriate design of the leaf springs.

(2) *Open Boundary Conditions.* All blocks have four neighbors. The system can be thought of as surrounded by an external frame to which the outer column and row of blocks are attached by springs of strength K_1 and K_2, respectively. No matter how one construes the motion of the surrounding rigid frame, the load of the boundary layer of blocks attached to the frame will be different from the load of the blocks in the bulk. One can think of the frame as either static – attached to the stationary plate beneath the sliding blocks – or as attached to the moving plate above the blocks. In either case, the tearing and shearing of the boundary blocks will be different from the forces exerted in the bulk. This implies that when one updates the boundary sites one should use an α different from the one used in the bulk, yet OFC did not do this. What is disheartening is that Leung et al. (1997) found that the SOC behavior is lost if a different value for α is used for the boundary sites; that is, SOC is lost if one includes the surrounding frame in a realistic way in the update algorithm. In this case, the model develops into a periodic state.

(3) *Periodic Boundary Conditions.* We are accustomed, in equilibrium statistical mechanics, to model infinite systems by assuming periodic boundary conditions for the models we simulate on computers. The pleasant aspect of periodic boundary conditions is that they are like having no boundary at all; all sites are equivalent. For some SOC models, periodic boundary conditions have a detrimental effect on the criticality of the model. In the case of the OFC model, the critical state is replaced by a periodic time evolution, as mentioned in Section 4.3.2.

Let us finally make a few comments concerning the relation between some quantities measured numerically in OFC simulations and some characteristics of earthquakes (see Section 3.6).

The Gutenberg–Richter Law. It is natural to relate the power law distribution of avalanches in the OFC model to the size distribution measured for earthquakes. The exponent τ in the OFC model varies over a rather broad range when the conservation parameter α varies between near zero and 1/4. The variation in τ includes the range $\tau \in [0.8, 2]$ observed for earthquakes occurring along faults at different geographical locations. The α-dependence of τ in the OFC model suggests that the variation observed in the exponent for real earthquakes is related to geographical variations in the strain and the elastic properties of the earth's crust.

The Omori Law. The $1/t$ decay in the excess number for large avalanches found in the OFC model may be related to the Omori law for the frequency of aftershocks following a large earthquake (see Section 3.6). There are, however, important differences. The Omori law is for the short-time behavior of the frequency of (small) aftershocks following a large earthquake, whereas the $1/t$ decay observed in the OFC model is for the excess frequency of *large* avalanches and therefore applies to long-time correlations. Moreover, the number of earthquake aftershocks decays like $1/t^a$ with a in the neighborhood of, but not exactly equal to, unity. Nevertheless, it is certainly conceivable that the Omori law is caused by the correlations generated during stick-slip dynamics in a way similar to the dynamics of the OFC model.

4.4 Lattice Gas

In this section we discuss fluctuations in the density of particles in a deterministic lattice gas. The model is an example of a many-body system that enters a state characterized by power laws. The system needs no tuning, and the behavior is essentially the same for open as for periodic boundaries (Jensen 1990; Fiig and Jensen 1993). We will first define the model and then consider some of its properties; we will also discuss possible relations between the model and real physical systems.

4.4.1 Definition of the Lattice Gas Model

The model is defined on a square lattice in d dimensions. Each cell of the lattice contains either one particle or no particle; double occupancy is not allowed.

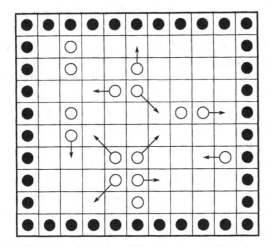

Figure 4.11. Examples of particle updates in the lattice gas model.

Neighbor particles repel each other by a unit force. The total force $\mathbf{F}$ on each particle is calculated. An update consists of displacing each particle from its position $\mathbf{r}_0$ to a new position $\mathbf{r}_1$, where $\mathbf{r}_1 = \mathbf{r}_0 + \mathbf{n}$. The displacement vector $\mathbf{n}$ is given by

$$\mathbf{n} = (n_1, n_2, \ldots, n_d) = ([F_1/F], [F_2/F], \ldots, [F_d/F]), \qquad (4.47)$$

where $F = |\mathbf{F}|$ and $[t]$ denotes the integer nearest to the real number t. The move is performed only when the site at $\mathbf{r}_1$ is unoccupied. Clearly, conflicts arise when two or more particles are to move onto the same lattice cell. In this case, the particle subject to the largest force wins and is moved onto the desired cell; the conflicting particles are not moved. Finally, if two or more particles – all subject to forces of equal strength – want to move onto the same site then none of the particles are moved. Figure 4.11 shows some examples of how the lattice is updated.

Open as well as periodic boundary conditions are considered. In the case of open boundaries, communication with the surroundings is established as follows. The lattice is surrounded by a set of stationary particles (see Figure 4.11). This construction ensures that particles in the boundary sites of the lattice, say $\mathbf{r} = (1, x_2, \ldots, x_d)$ or $\mathbf{r} = (x_1, 1, x_3, \ldots, x_d)$, feel a force toward the interior of the lattice. At the end of every update, the particles positioned in the boundary (next to the stationary particles) are annihilated. The first action in the next time step consists of visiting all the boundary sites. A particle is now placed in a boundary site with probability p. When this has been done the lattice is updated simultaneously. Some of the particles in the boundary site may then leave the boundary site and move toward the interior of the lattice if they are subject to a net force pointing toward an empty cell. Other particles may, as a consequence of the update, be moved onto the boundary and will then be annihilated at the end of the time step. In this way, the system exchanges particles with an exterior reservoir. The total number of particles in the system, $N(t)$, is therefore *not* conserved but fluctuates in time. The temporal fluctuations of $N(t)$ are of particular interest.

Instead of the open boundary, one can implement the usual periodic boundary condition. In this case the total number of particles on the system is obviously constant in time. The temporal fluctuations in the number of particles within a subvolume of the lattice is now studied. The periodic system can enter a static state when the density of particles is less than $1/2$, whereupon the particles can be arranged in configurations such that no particle has any nearest neighbor and therefore all forces are equal to zero. Numerically it is found that, when started from a random configuration, the system is unable to find the static configuration when the density is larger than about 0.3. The implementation of the algorithm as a computer code is discussed in Appendix B.

4.4.2 Properties of Lattice Gas

We will discuss the temporal as well as the spatial features of the (two-dimensional) model. Let us begin with the temporal fluctuations in the particle number $N(t)$. The power spectrum $S(f)$ of $N(t)$ is an accessible measure of the correlations in $N(t)$ (see Sections 2.3, 4.2.5, and Appendix D). Numerically, one finds that

$$S(f) \sim \begin{cases} \text{constant} & \text{if } f < f_{\text{cr}}, \\ 1/f^{\beta} & \text{if } f > f_{\text{cr}}, \end{cases} \qquad (4.48)$$

where $\beta \simeq 1$. This behavior is found for the open system when the driving frequency p is larger than about $p_{\text{cr}} \simeq 0.01$ and for the periodic system whenever the density is sufficiently large to inhibit a static state. The crossover frequency scales as $f_{\text{cr}} \sim 1/L^2$, where L is the linear extent of the lattice or the linear size of the considered subvolume in the case of periodic boundary conditions. This scaling suggests that $1/f_{\text{cr}}$ is determined by the diffusion time across the system. For $p < p_{\text{cr}}$, the dynamical connectivity across the system is lost, and the motion of the particles becomes essentially confined to the vicinity of the open edge. The result is that the power spectrum changes form to a power law with $\beta \simeq 2$.

Exercise. Simulate the model in three dimensions.

The individual particles are indeed found to behave as ordinary random walkers, as expected of diffusing particles. We consider the "lifetime distribution" $P(T)$ and the time dependence of the square displacement $R^2(t)$. We define the lifetime T of a particle as the number of time steps completed from the instant a particle enters the lattice (or, for periodic boundary conditions, the subvolume being probed) until it exits over some part of the boundary. The distribution of lifetimes, or resident times, scales as $P(T) \sim T^{-\alpha}$ ($\alpha \simeq 3/2$) for $T < T_{cr}$ and goes rapidly to zero for $T > T_{cr}$. The cutoff time increases as $T_{cr} \sim L^2$; that is, we again have diffusive behavior. The exponent $\alpha = 3/2$ is identical to the exponent for the return-time distribution of independent random walkers (Fogedby et al. 1991). The return time is defined in the following way. Consider a random walker in d dimensions. Divide the space into two parts by a hyperplane. The *return time* is the time between successive crossings of the hyperplane – namely, the time it takes to return to the hyperplane. This distribution is a power law with exponent $-3/2$.

The square displacement $R^2(t)$ of particles is easily measured during a simulation. The position $\mathbf{r}(t_0)$ at time t_0 is recorded. The system is then updated t time steps, and the position $\mathbf{r}(t_0 + t)$ is once again recorded. The square displacement is calculated from

$$R^2(t) = \langle [\mathbf{r}(t_0 + t) - \mathbf{r}(t_0)]^2 \rangle, \qquad (4.49)$$

where the average is over t_0 and particles. The time dependence is $R^2(t) \sim t$, that is, ordinary diffusion.

We conclude that, by following an individual particle, one sees this particle bumping into other particles every so often. All these collisions make the particle perform an erratic motion characteristic of random walks. However, there is more to the motion of the particles than this. We previously considered fluctuations in the total number of particles $N(t)$ inside a specified area. For *independent* uncorrelated random walkers the power spectrum of $N(t)$ behaves like $1/f^{3/2}$. This is most easily seen in the following way. One can think of $N(t)$ as the sum of time signals

$$N(t) = \sum_i n_i(t), \qquad (4.50)$$

where $n_i(t) = 1$ whenever particle i is inside the area being monitored and $n_i(t) = 0$ when the particle is outside this area. The signal $t \rightarrow n_i(t)$ consists of trains of boxes of varying length T distributed according to $P(T) \sim T^{-\alpha}$. If the motion of the particles were uncorrelated then we would have $S(f) \sim f^{-\beta}$ with $\beta = 3 - \alpha$, as discussed in Section 4.2.5, and therefore $\beta = 3/2$ and not the measured value $\beta \simeq 1$. The random walker exponents $\alpha = \beta =$

3/2 are indeed obtained when the deterministic dynamics considered here is replaced by stochastic Monte Carlo dynamics (Andersen, Jensen, and Mouritsen 1991). The power spectrum measured for the deterministic lattice gas has an exponent $\beta \simeq 1$. From this we conclude that, although the particles behave like random walkers individually, there are nevertheless important correlations between them. The surprising finding is that these correlations are not destroyed by interparticle collisions.

Exercise. Simulate the two-dimensional model, using the following stochastic random walk dynamics. Visit the particles on the lattice in a random sequence. Choose at random one of the eight neighbor sites and move the particle to that neighbor site if the site is empty; otherwise, leave the particle where it is. Measure the power spectrum of the $N(t)$ and measure the lifetime probability density $P(T)$ as well as the square displacement $R^2(t)$.

The density fluctuations, as characterized by the power spectrum of $N(t)$, exhibit rather subtle long-time correlations. The only time scale present is the diffusion time for traversing the system. Are the spatial characteristics interesting and scale-invariant? This depends on the quantity considered. The density–density correlation function is not very interesting; it is essentially structure-free (Fiig and Jensen 1993). But the spatial arrangement of the energy dissipation exhibits interesting properties (Jensen 1990). The dynamics of the model corresponds to overdamped dynamics for which the velocity is proportional to the acting force. Thus, we can imagine that energy is dissipated every time a particle moves. One can follow the motion of the particles from one time step to the next. Suppose we mark all the sites from which particles leave at time step t and mark, in addition, all the sites onto which these particles move at time $t + 1$. The marked sites form clusters of dynamically connected sites. The size of clusters varies. The probability density of cluster sizes follows a power law $P(s) \sim s^{-\gamma}$, where $\gamma \simeq 2$.

4.4.3 The Lesson of Lattice Gas

What do we learn from this model? It is an example of a system that organizes itself into a state without characteristic scales, except for those induced by the finite size of the system. The model also shows that $1/f$ spectra can arise in this state. It is certainly instructive to notice that the $1/f$ spectra of the model do not arise as a simple linear superposition of independent time signals of varying duration (see Section 4.2 and Appendix D). Rather, the power spectrum of the number density $N(t)$ is a consequence of subtle long-time correlations *between* the individual events constituting $N(t)$. I think all these features of the model make it reasonable to consider the lattice gas as an example of a SOC system. Is there a threshold in the model? The answer is Yes. A particle will move only if the resulting force $\mathbf{F}$ on that particle is of a magnitude $|\mathbf{F}| > F_{thr}$. We used $F_{thr} = 0$ in our previous discussion of the model, but it is obvious that we could just as well have chosen F_{thr} equal to, say, half the magnitude of the force acting between two nearest neighbor particles. The important point is that the activity in the system is caused by *interaction* between the particles. Isolated particles do not move – that is, no interaction means no activity.

It is worthwhile to examine the boundary condition. The probability p that a particle is introduced at the empty boundary sites acts as a driving rate. Obviously, the higher the value of p the more hectic is the activity in the edge layer of the system. The average density of particles, $n(p)$, does depend on p in a way that is roughly given by the solution to

$$p(1 - n) = n^2(1 - p). \tag{4.51}$$

This equation is approximately the condition for balance between the number of particles entering the system and the number of particles leaving the system. A particle is placed at an empty edge site at the beginning of each time step with probability p. It will move into the interior of the system if its neighbor site is empty, which happens with probability $(1 - n)$. A particle located next to the edge sites will leave the system with probability $n(1 - p)n$. The first factor n is the probability that a site next to the edge is occupied; the factor $(1 - p)$ is the probability that the edge site next to this occupied site remains empty after the introduction of particles into the edge sites during the first stage of the update. Finally, the last factor n is the probability that the particle under consideration has a neighbor next to it that can push the particle off the lattice. See Figure 4.12 for a comparison between the analytic expression for $n(p)$ derived from (4.51) and the simulation result.

Exercises. See Figure 4.12.
(a) Explain the deviation between the simulation results and the analytic expression for $n(p)$.
(b) Include higher order processes (two or more neighbors, etc.) in (4.51) for $n(p)$.

Notwithstanding the dependence of $n(p)$ upon p, the system organizes itself into a state with the same type of $1/f$ fluctuations for a broad range of p-values. A $1/f$ spectrum is found for $p \in [0.01, 1]$, approximately. Similarly, the periodic system exhibits the same fluctuation spectrum for any density larger than

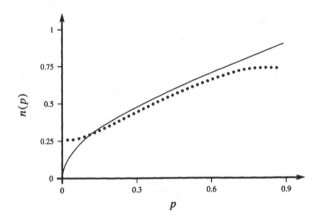

Figure 4.12. Density of particles on the lattice as a function of the introduction probability p. The solid line is the solution to (4.51); the dots represent data points from the simulations reported in Fiig and Jensen (1993).

about 0.3. Thus, the scale-invariant state selected by the system is robust with respect to changes in particle density.

Perhaps the most surprising lesson from the study of this model is its sensitivity to the definition of the update algorithm. Given that individual particle motion is characterized by the standard properties of random walkers, one might have imagined that the fluctuation spectrum of the total number of particles would be insensitive to the introduction of a stochastic element in the updating algorithm. However, a stochastic element in the update changes the power spectrum from $S(f) \sim 1/f$ to $S(f) \sim 1/f^{3/2}$, where the latter is the behavior of independent random walkers.

4.4.4 Physical Relevance of the Lattice Gas Model

The deterministic lattice gas model may be viewed as a description of the flux noise experiment performed some years ago by Yeh and Kao (1984). Their experiment is concerned with the onset of motion of vortices in a superconducting sample (see Section 3.4 for a discussion of superconducting vortices and vortex pinning).

A loop consisting of thin superconducting wire is glued onto one of the faces of the sample (see Figure 4.13). This loop is connected to a SQUID, an instrument that measures the exact number of vortices threading the area of the sample enclosed by the loop. An external current is applied to the sample. In the

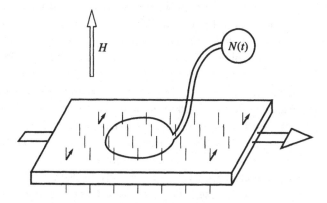

Figure 4.13. The experiment on fluctuations in vortex density in the vicinity of the depinning threshold. The large horizontal arrow indicates the direction of the electrical current driven through the sample. The four small thick arrows indicate the direction of the Lorentz force induced by the applied current. The vertical arrow indicates the direction of the applied magnetic field.

current regime just above the depinning current (see Section 3.4), where the vortex system starts gradually to move, Yeh and Kao found that the number of vortices within the loop area fluctuates with a $1/f$ power spectrum. The experiment is performed at low enough temperatures to make the vortex–vortex interaction energy dominate over the thermal energy. The thin superconducting wire of the pickup loop acts as a weak barrier that the vortices must overcome in order to enter or leave the area probed. It is natural to imagine that the motion over this barrier is thermally assisted and therefore contains a degree of randomness. On the other hand, the motion inside the enclosed area is probably governed mainly by the deterministic intervortex forces. This interpretation of the experiment maps it, in qualitative terms, directly to the deterministic lattice gas. Other features of the lattice gas model also fit the behavior of this experiment. When the current is increased well above the depinning threshold, the power spectrum crosses over from $1/f$ behavior to $1/f^2$ behavior. The same is seen with the lattice gas and, indeed, also follows from a description in terms of the diffusion equation. See (5.18) and Grinstein, Hwa, and Jensen (1992).

4.5 Forest Fires

In this section we describe a model developed by Drossel and Schwabl (1992), known as the *forest fire model*. The degree of relationship to real forest fires remains unknown, but the model has been used with some success to model the spreading of measles in the population of the island of Bornholm and of the Faroe Islands (Rhodes and Anderson 1996; Rhodes, Jensen, and Anderson 1997). To be consistent with the literature we will use the terminology of trees and fires. The harm in doing so is no greater than in using sandpile language for the BTW cellular automata (see Section 4.2).

We include the model in our discussion for several reasons. First, the model is simple and elegant and may indeed throw some light on the burstlike temporal fluctuations observed in epidemiology. Second, numerical simulations show that the model is critical when driven in a certain limit. Third, a neat mean field theory exists for the model (Christensen, Flyvbjerg, and Olami 1993). We describe the mean field theory in Section 5.2.4. Finally, the Rome group has analyzed the forest fire model by use of their "dynamically driven renormalization group" technique (Loreto et al. 1995). This analysis, which we discuss in Section 5.5, shows that the model is indeed controlled by a critical point. However, the critical point is *repulsive*. Hence the model needs some kind of fine tuning in order to exhibit critical fluctuations; that is, the model is not *self*-organized critical. But the forest fire model can be tuned to criticality in a way similar to the tuning of a thermodynamic equilibrium system to the critical temperature. The aspects mentioned here make the model excellent as a tool for learning something about the conditions needed for self-organization to criticality and also as a target for testing formalisms.

4.5.1 Definition of a Critical Forest Fire Model

The model is defined on a d-dimensional cubic lattice. Each site of the lattice can be in one of three different states: the site can be empty; the site can contain a green tree; the site can contain a burning tree. The lattice is updated in parallel according to the following algorithm.

(1) A site occupied by a burning tree becomes an empty site in the succeeding time step.
(2) A green tree becomes a burning tree if one or more of its nearest neighbor sites contain a burning tree.
(3) An empty site becomes occupied by a green tree with probability p (the growth rate) in each time step.
(4) A green tree that is not neighbor to burning sites catches fire spontaneously with probability f (the lightning rate) in each time step.

Periodic boundary conditions are assumed, and the initial configuration can, for instance, be a random configuration of green trees and empty sites.

Rule (4) turns out to be crucial for the critical behavior. If the lattice is updated according to the first three rules only, the evolution becomes periodic rather than critical. System-spanning spiral-like fire fronts traverse the system with a period proportional to $1/p$ (Grassberger and Kantz 1991; Mossner, Drossel, and Schwabl 1992). A characteristic scale develops when the spontaneous ignition is left out. This can be understood in the following way. Small clusters of green trees will inevitably become large clusters, since nothing limits their size except becoming connected to a cluster of trees that contains a burning front. A characteristic size of stable tree clusters can be estimated by balancing the growth against the number of burning trees. For a cluster size to be stable, the number N_b of trees burning away in each time step must be balanced by the number of new green trees N_g grown along the perimeter of the cluster during a time step. New trees grow along the empty sites along the perimeter of the considered cluster. Let N_p denote the number of sites along the perimeter. We then have $N_g \sim pN_p$. Because sites neighbor to burning sites catch fire with probability 1, the number of burning sites N_b will in each time step be of order 1 or larger. Thus, balance between N_b and N_g implies that N_p must increase like $1/p$ when $p \to 0$. In this way we see that a characteristic scale given by $1/p$ exists.

Exercise. Why is critical behavior a possibility only in the limit $p \to 0$?

Numerical simulations as well as analytical analysis (Sections 4.5.2 and 5.5.4) show that rule (4) in the updating algorithm secures the critical behavior in the double limit $p \to 0$ and $f/p \to 0$. The condition that $f \ll p$ is important. If the ignition probability is too high then large-scale structures are never formed. Clusters of green trees simply burn down before they manage to grow large. The necessity of random spontaneous ignition reminds us of the behavior in the OFC model (Section 4.3), where the synchronized periodic behavior found for periodic boundary conditions is broken by the inhomogeneity introduced by an open boundary. Random ignition appears to have a qualitatively similar effect in the forest fire model.

4.5.2 Simulation Results for the Forest Fire Model

Simulations of the model in dimensions 1–8 confirm its critical behavior (Clar, Drossel, and Schwabl 1994). The probability $P(s)$ that a cluster of green trees contains s trees exhibits the power law behavior $P(s) \sim s^{-\tau}$, where $\tau = 1$ in dimension 1 and increases to $\tau = 1.5$ in dimension 6. The clusters of green trees are fractal objects in dimensions larger than 2. This is seen from the scaling of the radius of gyration $R(s)$. The *radius of gyration* for a cluster of size s is defined as the mean distance of the trees from the center of mass of the cluster. That is,

$$R(s) = \frac{1}{s} \sum_{i=1}^{s} |\mathbf{r}_i - \langle \mathbf{r}_i \rangle|. \tag{4.52}$$

The simulations show that $R(s) \sim s^{1/\mu}$. This scaling indicates a fractal structure when $\mu < d$. The numerics show this to be the case for $d > 2$ and perhaps also for $d = 2$ (Grassberger 1993; Clar et al. 1994). The simulations find that the scaling exponents change with dimension for $d < 6$ and thereafter remain independent of dimension (for $d \geq 6$). This suggests that the model's upper critical dimension $\equiv 6$. This is in agreement with analytic mean field arguments involving a mapping of the forest fire model onto percolation (see Section 5.2.4).

4.5.3 Physical Relevance of the Forest Fire Model

The temporal variation in the recorded number of measles cases is very accurately registered on the Danish island of Bornholm as well as on the Faroe Islands. The number of cases fluctuates greatly. Bursts of varying sizes in the number of infected individuals are followed by longer or shorter periods of quiescence. These islands are rather isolated, so the development of an epidemic burst can perhaps be thought of as driven by more-or-less internal interaction among members of the population. New outbursts are probably to a large extent induced at random by visits from the outside. If this is the case then the dynamics of the spreading of measles resembles the dynamics of the forest fire model.

In fact, Rhodes and Anderson (1996) and Rhodes et al. (1997) have found an encouraging similarity between temporal variation in the number of fires predicted by the forest fire model and the temporal records of the incidence of measles on these islands. One reservation that springs to mind is, of course, that in the forest fire model the fire spreads with certainty upon contact, whereas transmission of measles occurs only with some probability $p < 1$. Children do not *necessarily* develop measles just because they visit the neighbor's son who happens to be in bed with spots all over his body. One promising aspect is that the critical exponents of the forest fire model are found to remain unchanged if, in rule (2) of the defining algorithm, we introduce a spreading probability that is less than unity (Drossel and Schwabl 1993).

4.6 Extremum Dynamics

The threshold dynamics of the sandpile and earthquake models discussed in the beginning of this chapter inspire a new way of summarizing the dynamics of certain systems. In systems where the dynamical evolution is a struggle against various types of thresholds or barriers, the action will predominantly occur where the net barrier is the smallest. The effective local barrier is determined by the difference between the total load applied locally and the local barriers; the location of action is found through a global search for the smallest local barrier. In the earthquake model, for instance, the site with the largest force (and therefore the smallest distance to the threshold for slipping) is identified. The global load on the system is increased precisely enough to make this site critical, and then the system is relaxed. The action is induced at the site of extremal strain.

In this section we discuss two fascinating examples of what we may call extremum dynamics. The first is a model inspired by the motion of an elastic string or a growing interface through a disordered environment (Sneppen 1992; Zaitsev 1992). The second model is known as the Bak–Sneppen model of evolution (Bak and Sneppen 1993), which is presumed to have some (abstract) relevance to biological evolution (see Section 3.7). As hard as it is to bring the evolution model to a weighty test, so has it been debated to what extent the Bak–Sneppen model is of relevance to biological evolution. We needn't worry too much here about the biological justification of the Bak–Sneppen model, which grew out of the Sneppen model of interface depinning. The depinning model is probably qualitatively related to the physics of interface depinning, although it has some shortcomings (Roux and Hansen 1994; Jensen 1995).

The strength of these abstract extremum models is that they are extremely simple to define. They are exceptionally easy to simulate, their mean field theory is affordable, and one might even be able to discuss them analytically beyond mean field level (Paczuski, Maslov, and Bak 1994, 1996). The shortcomings are that such models are formulated on purely heuristic grounds, which makes it difficult to assess the relevance of predictions extracted from them. In fact, one could make this criticism against most of the algorithms put forward in the field of SOC – a situation characteristic of emerging disciplines. How long did it take before it became clear that the simple Ising spin model actually does describe aspects of *real* magnetic systems?

After all these words of caution, let us now turn to the definition of the models.

4.6.1 The Model of Interface Growth in a Random Medium

Consider a fluid being pumped into a porous rock, or a flux line (see Section 3.4) being forced through a pinning potential of an inhomogeneous superconductor, or for that matter the jerky motion of the triple line† every time we pull the spoon from our cup of coffee. In all cases the elastic medium (the front of

---

† The "triple line" is the line along the spoon where air, liquid, and solid all meet.

the liquid, the magnetic flux line, or the line of interface between the coffee, the spoon, and the air) is subject to a global drive that fights against the friction or pinning forces, which in turn are due to the interaction between the elastic medium and the environment through (or over) which it is being forced. The motion consists of a competition between the global drive, the elastic restoring forces of the interface, and the pinning forces exerted by the surrounding medium. In the example of the liquid being pumped through a rock (as when extracting oil from underground wells), the driving force is set up by the pressure exerted on the liquid being pumped through the rock. The restoring forces arise from the surface tension, and the pinning forces from the viscous friction from the internal wall of the rock. The front of the liquid moves ahead in a given time instant at the point of least resistance.

For simplicity we describe the Sneppen (1992) interface growth model in one dimension; the generalization to higher dimensions is straightforward. Consider the function $x \to h(x)$ for $x = 1, 2, \ldots, L$. We can think of $h(x)$ as describing the height of the interface. This representation obviously excludes overhangs, a limitation that is worthwhile to bear in mind. The interface is being acted on by random pinning forces, which are described by the function $\eta(x, h(x))$ defined in the x–h plane. At each point (x, h), the value $\eta(x, h)$ is drawn from a random distribution $P(\eta)$ of variance σ^2. The values of $\eta(x, h)$ are assumed to be delta-correlated; that is,

$$\langle \eta(x, h)\eta(x', h') \rangle = \sigma^2 \delta(x - x')\delta(h - h'). \tag{4.53}$$

The applied drive of the interface is incorporated through an updating procedure as follows.

(1) The site $x_{\min}$ with the smallest value of $\eta(x, h(x))$ is localized.
(2) The height at $x_{\min}$ is increased by one unit $h(x_{\min}) \to h(x_{\min}) + 1$.

The move forward by the interface at position $x_{\min}$ will increase the pull at the neighboring sites. The extra pull may enable the interface to move forward in the vicinity of the extremum site $x_{\min}$. This aspect is included in the dynamics of the Sneppen model by the following step.

(3) The neighboring sites $\ldots, x_{\min} \pm 1, x_{\min} \pm 2, \ldots$ are successively adjusted upward by the increment $h(x_{\min} \pm k) \to h(x_{\min} \pm k) + 1$ for $k = 1, 2, \ldots$ until all the local slopes assume values $|h(x + 1) - h(x)| \le 1$. The increment is first applied to the sites $x_{\min} - 1$ and $x_{\min} + 1$, then to the sites $x_{\min} - 2$ and $x_{\min} + 2$, and so on if needed.

Step (3) leads to avalanches of activity spreading out from the minimal site $x_{\min}$. The procedure is started over again by returning to step (1) once all local slopes have yet again assumed values less than or equal to unity.

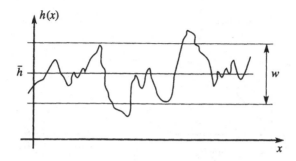

Figure 4.14. Width and average height of a growing interface.

Notice that when we implement this algorithm on a computer we do not need to specify and store the random field $\eta(x, h)$ beforehand. Because no correlations exist between $\eta(x, h)$ at different positions (x, h), one can simply draw upon random numbers as the interface moves ahead. Notice also that the dynamics is solely determined by the extremal properties of the random field $\eta(x, h(x))$. A high-quality random number generator is therefore extremely important for meaningful simulations of the model.

The dynamics defined in steps (1)–(3) leads to a jerky motion where large parts of the interface remain static for long periods and then suddenly make large jumps forward during the course of relatively few time steps. At least two aspects of the motion are interesting. One is the nature of these avalanches of activity; another (related) aspect is the behavior of the width of the interface (see Figure 4.14). We define the *width* of the interface as

$$w(t, L) = \left[\frac{1}{L} \sum_{i=1}^{L} (h(x, t) - \bar{h})^2 \right]^{1/2} \tag{4.54}$$

(Barabási and Stanley 1995), where $\bar{h}$ is the average height at time t. The dependence of the width on time is given by

$$w(t, L) \sim \begin{cases} t^{\beta} & \text{for } t < t_{\text{cr}}, \\ w_{\infty}(L) & \text{for } t > t_{\text{cr}}; \end{cases} \tag{4.55}$$

see also Figure 4.15. The dependence of the saturation level $w_{\infty}(L)$ on the length of the interface is given by

$$w_{\infty}(L) \sim L^{\chi}. \tag{4.56}$$

The value of the roughness exponents $\beta = 0.9 \pm 0.1$ and $\chi = 0.633 \pm 0.001$ (Sneppen 1992) have created much excitement, since they differ significantly

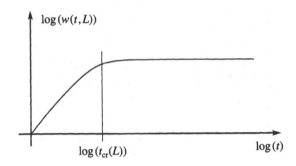

Figure 4.15. Time dependence of the width of a growing interface.

from the predictions $\beta = 1/3$ and $\chi = 1/2$ of the leading theory of interface growth, the KPZ equation (Kardar, Parisi, and Zhang 1986). The increased values of the roughness exponents are caused by the presence of static quenched-in disorder, which is not included in the KPZ equation. The exponent χ has actually been found to exceed unity in simulations of a linear elastic string dragged through a random potential (Roux and Hansen 1994; Jensen 1995).

The Sneppen model is in some sense an example of a self-organized critical system. The width of the interface exhibits power law behavior, as does the distribution of the avalanches released during the jerky forward motion of the interface. The *size* of an avalanche is defined as the area of the region in the *x–h* plane covered when part of the interface jumps ahead (see Figure 4.16). The probability density of avalanche sizes has the form $P(s) \sim s^{-\tau}$, where $\tau \simeq 1.26$ (Paczuski et al. 1996).

As defined in the three-step algorithm, the dynamics of the model contains no apparent tuning. There is, however, some implicit tuning assumed. The algorithm presupposes that the center of mass of the interface moves more slowly than any process involved in the relaxation of the strain along the interface (step (3)). In fact, the algorithm is equivalent to the limit of a constant but infinitesimally small velocity of the center of mass. If the center of mass were moving with finite velocity then one would not observe large portions of the interface being static for long periods of time. In the limit of a constant but slow center-of-mass velocity, the total applied force is constantly adjusted to precisely counterbalance the pinning force exerted by the surrounding media (Jensen 1995). This mode of driving certainly is physically realizable. In some experiments it is indeed simple to impose this velocity – say, by pumping at a constant rate the liquid that is being forced through the porous medium. It must be kept in mind that prescribing a constant velocity leads to a physical

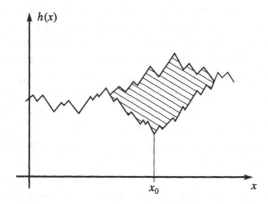

Figure 4.16. Structure of an avalanche in the Sneppen interface model.

situation different from the one in which a constant applied force is prescribed. The force correlations and fluctuations in the velocity-driven experiment are bound to be different and larger than in the force-driven experiment.

The pinning force given by $\eta(x, h)$ has an interesting distribution in the stationary state of the model. At the beginning of the simulation of the motion of the interface, the probability density $P_i(\eta(x, h(x)))$ of $\eta(x, h(x))$ values along the string is equal to the probability density $P(\eta)$ from which the variables η are drawn. The form of $P_i(\eta, h(x))$ changes during the iteration of the dynamics, and $P_i(\eta, h(x))$ asymptotically approaches a distribution that is uniform on $[\eta_c, 1]$ (where $\eta_c \simeq 0.4614$; see Pazuski et al. 1996) if the distribution $P(\eta)$ is uniform on $[0, 1]$. We shall return to this aspect of extremum models when discussing the evolution model in Section 4.6.2.

Exercises. (1) Study the asymptotic form of the distribution $P_i(\eta(x, h(x)))$ for distributions $P(\eta)$ different from uniform distribution.

(2) Study the influence of using nonuniform distributions of $P(\eta)$ on the way the stationary state is approached.

4.6.2 The Evolution Model

We consider only the one-dimensional version of the model, although a generalization to arbitrary dimension can be made immediately. Consider a one-dimensional lattice of L sites. We assume periodic boundary conditions. To each lattice site x we allocate a number $B(x) \in [0, 1)$. In the language of evolution, $B(x)$ is called a *fitness barrier* (Bak and Sneppen 1993). The species x

will have to overcome a certain barrier in order to evolve toward higher fitness. Fitness is a relative quantity that depends on the properties of the surrounding species. The least fit species is subject to the smallest fitness barrier. In other words, the lesser is one's fitness, the more likely it is that this fitness will be improved by a random mutation. These airy considerations suggest the following updating algorithm for the model. Locate the position x_{min} of the site with the smallest barrier $B(x)$. The species x_{min} undergoes a random mutation that changes its barrier at random. This is described by replacing the current value $B(x_{min})$ by a new random number chosen with uniform probability on the interval $[0, 1)$. The change of the properties of species x_{min} induces a change in the life of species in contact with species x_{min}. To mimic this, one also replaces the barrier $B(x_{min} - 1)$ and $B(x_{min} + 1)$ by new random values from the interval $[0, 1)$. The update algorithm may be summarized as follows.

(1) Locate site x_{min} with the smallest $B(x)$ value.
(2) Perform the substitutions

$$B(x_{min} - 1) \to u_1,$$
$$B(x_{min}) \to u_2, \tag{4.57}$$
$$B(x_{min} + 1) \to u_3,$$

where $u_i \in [0, 1]$ are random numbers drawn with uniform density.

A computer code for this algorithm is discussed in Appendix C.

This simple dynamical rule leads to some interesting behavior of the model. The distribution of barriers develops toward a step function, as shown in Figure 4.17. In the limit of infinite system size, the distribution is characterized by a single parameter B_c. For $B < B_c$ the distribution is zero, $P(B) = 0$. When $B > B_c$, the distribution assumes the constant value $P(B) = 1/(1 - B_c)$ determined by the normalization condition. The usual two-point correlation function exhibits algebraic decay, which is characteristic for a critical system. One finds

$$G(x) = \langle B(x')B(x + x')\rangle_{x'} - \langle B(x')\rangle^2 \sim x^{-\eta}e^{-x/x_0}, \tag{4.58}$$

where $\eta \simeq 0.7$ and where $x_0 \to \infty$ when $L \to \infty$ (see Datta, Gilhøj, and Jensen 1997).

It is also possible to define burstlike avalanches in this model. Let $B_{min}(t)$ denote the smallest of the B-values present in the model at time t. As the model is updated, the value $B_{min}(t)$ changes. Let B_0 denote some fixed B-value. One can now measure the time interval between successive updates where $B_{min}(t)$ crosses the value B_0. Assume that $B_{min}(t) > B_0$ but that $B_{min}(t') < B_0$ for $t < t' < T$ and $B_{min}(T + 1) > B_0$. We then say that an avalanche of duration (or

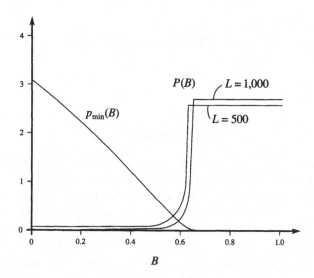

Figure 4.17. Probability density for the fitness barrier in the Bak–Sneppen model. The size dependence of $p_{\min}(B)$ (the distribution of the minimum barriers) is not visible on the scale of the figure.

lifetime) T has taken place. If $B_0 = B_c$ one finds that the probability density of avalanche lifetimes $P(T)$ exhibits the power law behavior $P(T) \sim 1/T^{\alpha}$, where $\alpha \simeq 1.0$ (see Bak and Sneppen 1993).

The critical nature of the state established by the repetitive action of the updating algorithm can be seen in a number of quantities. One can, for instance study the distribution of distances $\delta x = |x_{\min}(t+1) - x_{\min}(t)|$ between consecutive updates. The probability density is found to behave algebraically according to $P(\delta x) \sim \delta x^{-a}$ with $a \simeq 3.1$ (Bak and Sneppen 1993). This algebraic form of $P(\delta x)$ is of course related to correlations between the updated sites. We shall return to this point in Section 5.2.5.

5

The Search for a Formalism

5.1 Introduction

Self-organized criticality was introduced in 1987 by Bak, Tang, and Wiesenfeld (BTW), who employed some appealing yet heuristic handwaving arguments. To substantiate the hypothesis, computer simulations of a simple algorithm – inspired by the avalanches induced when one plays with a pile of sand – were presented. Substantial analytic understanding was lacking for some time. Soon, however, standard mathematical tools were being applied to the new set of models. Statistical mechanics traditionally makes use of a combination of approaches, one of which consists of defining models that have a structure that allows an exact calculation of specific quantities. The art is to formulate a model of the right degree of complexity. One wants a model with sufficient structure to contain nonobvious behavior, but the model should not be so complicated that analytic approaches cannot be carried through. This last property obviously depends strongly on who is going to perform the analysis. The mathematical power of Deepak Dhar and his co-workers made it possible for them to solve an only slightly altered version of the original BTW cellular automata (Dhar and Ramaswamy 1989; Dhar 1990). After Dhar's work it was clear that, at least in some cases, the observed critical behavior was not merely an artefact of simulations on too-small systems. We shall outline the approach developed by Dhar and co-workers in Section 5.3.

Despite their undeniable beauty, the exact solutions have one drawback: the specific mathematics tends to be tailored to the details of the solved model. This means that generalization to other, similar models is often not possible. In equilibrium statistical dynamics, as well as in the study of driven systems of many degrees of freedom, various approximate approaches have been developed in order to calculate in a systematic way the large-scale, long-time behavior of the models. One powerful method has been to consider the models at a coarse-grained level. This approach is the same as that used in hydrodynamics. When one considers a macroscopic number of particles, it becomes intractable to follow the motion of the individual particles. Instead, densities $\rho(\mathbf{r})$ and velocity fields $v(\mathbf{r})$ are introduced that describe the average number of particles and their average velocity in a unit volume at position $\mathbf{r}$. Derivation of the effective coarse-grained equation of motion is not, in general, a rigorous procedure. One makes use of known conservation laws and symmetries of the problem to justify the form of the effective equation assumed to describe the large-scale behavior of the model under consideration. In Section 5.4 we shall discuss the difficulties of applying this approach to self-organized critical systems. Specific problems arise because of the sharp threshold involved in the dynamics. This threshold is represented in the equation of motion by a highly nonlinear and nonanalytic term.

If one can deduce a reliable set of coarse-grained equations of motion, it may then be possible to apply the renormalization group approach in order to analyze the large-scale behavior of the system. The renormalization method has for the last 25 years or so been the flagship among techniques applied by theoretical physicists. It has been used with success to analyze a wide variety of systems, including the quantum theory of high-energy physics, thermodynamic phase transitions, and the correlated motion of electrons in strong magnetic fields. This success has stimulated attempts to bring the renormalization group to bear on models of self-organized criticality. As we shall see, the strong nonlinearity of the problem makes the hydrodynamic or Langevin equation approach problematic. Moreover, it is difficult to identify quantities like the distribution of avalanche lifetimes and avalanche sizes in the Langevin treatment.

A different route has, however, been developed. This method resembles the real-space renormalization group approach of equilibrium statistical mechanics, and is known as the "dynamically driven renormalization group" (DDRG) method (although in the first papers it was called a "fixed-scale transformation"). We discuss this method in Section 5.5. The approach was first developed in order to calculate properties of fractal structures. One strength of the method is that the exponents of the various event distributions can be calculated in a direct way.

The virtue of the more rigorous and accurate methods is, of course, that they (may) produce reliable predictions. However, the mathematical complexity can be prohibitive. It is therefore useful to make a rough qualitative analysis of the considered systems – an analysis that (it is hoped) describes the gross features correctly even though it may be incorrect in some of the details. Mean field theory is such an approach. The next section is concerned with the salient aspects of this type of analysis.

5.2 Mean Field Theory

Mean field theory consists of an estimate of the "on-the-average" behavior of many interacting degrees of freedom. There are several different types of mean field theory. Their common feature is that the specific details of the surroundings are replaced by the typical average behavior. The theory by Weiss of the magnetic phase transition is a prototypical example of a mean field description. Weiss theory is detailed in Reif (1965, sec. 10.7) and in Feynman (1964, sec. II-36-6). To illustrate the use of mean field arguments when analyzing SOC models, we shall discuss the nature of the critical state in the sandpile model, the role of nonconservation in the earthquake model, the critical dynamics in the forest fire model, and the barrier distribution in the evolution model. (The details of all four models were given in Chapter 4.) The strength of mean field considerations is their simplicity, which allows one to develop a clear conceptual picture of the mechanics of the description. The oversimplification assumed in mean field theory is, at the same time, its greatest weakness. Of course, mean field calculations are not expected to be quantitatively accurate. The hope, though, is that its predictions are nonetheless qualitatively correct.

5.2.1 Sandpile Models

Consider a random neighbor version of the BTW sandpile cellular automaton (described in detail in Section 4.2). The effect of a random assignment of neighbors instead of the usual nearest neighbor arrangement is to destroy spatial correlations, as discussed in Section 4.3.3. We will in this sense consider the random neighbor model as a kind of mean field theory.

The model is defined as in Section 4.2.2 except that we chose $q_c = \alpha z_c$ new neighbors at random every time a site topples (Christensen and Olami 1993). Here q_c denotes the coordinate number of the site and α is the conservation level of the model, as we now explain. Each of the randomly chosen sites receives one unit of sand, $z_m \to z_m + 1$. The overcritcal site, say site i, loses z_c grains of sand, that is, $z_i \to z_i - z_c$. Notice that an element of dissipation has been introduced. Because only αz_c sites receive a grain of sand, the model is nonconservative except in the limit $\alpha \to 1$. We imagine that the model is driven by adding extra units of sand at random positions (this is rule R2 in Section 4.2.2).

It is illuminating to think of the evolution of avalanches in the model as a kind of branching process (Harris 1963). Figure 5.1 illustrates the treelike structure created by a branching process. The process starts at the top. Each node branches off a number n of branches. The probability that a node divides into $n = 0, 1, 2, \ldots$ branches is denoted by p_n. A toppling site can induce from zero to q_c new topplings. The temporal evolution of an avalanche can be considered as a branching process in which from zero to q_c new branches can

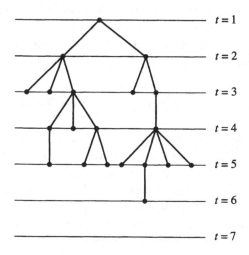

Figure 5.1. A tree generated by a branching process.

emerge. The average number of new branches is called the *branching ratio* $\sigma = \langle n \rangle$. The probability that a tree in a branching process of branching ratio σ develops s nodes before it dies is given by

$$P(s) \sim s^{-3/2} \exp\left(- \frac{s}{s_0(\sigma)} \right) \tag{5.1}$$

(Harris 1963). The exponential cutoff diverges when $\sigma \rightarrow 1^-$ according to $s_0(\sigma) \sim (1 - \sigma)^{-2}$. The tree size distribution is equal to the avalanche size distribution when we consider the sandpile as a branching process.

The branching probability is easily expressed in terms of the probability density $P(z)$. Here $P(z)$ denotes the probability that the dynamical variable z_i of an arbitrarily chosen site i assumes the value z, where z is an integer. A site becomes overcritical when $z_i \geq z_c$. Hence, the q_c grains distributed when a site topples will induce new topplings (or new branches) when they hit sites with z-values equal to $z_c - 1$. The probability that a toppling induces n new topplings is therefore given by

$$p_n = K_{q_c,n} P^n(z_c - 1)(1 - P(z_c - 1))^{q_c - 1} \tag{5.2}$$

(Christensen and Olami 1993), where $K_{i,j}$ denotes the binomial coefficient. The average branching ratio σ is given by

$$\sigma = \langle n \rangle = \sum_{n=0}^{q_c} n p_n = q_c P(z_c - 1). \tag{5.3}$$

The probabilities $P(z)$ for $z < z_c$ are easily determined from their master equation in the stationary state. The master equation expresses the change of $P(z)$ with time. Let $P(z, t)$ denote the probability that a site in the system assumes the value z at time t. The probability $P(z, t + 1)$ in the succeeding time step is given by

$$P(z, t + 1) = P(z, t) + \sum_{z'} [W_{z,z';t} P(z', t) - W_{z',z;t} P(z, t)]. \qquad (5.4)$$

The fraction of states that change their z-value from z' to z during a time step is denoted by $W_{z,z';t}$. The first term inside the brackets corresponds to sites changing their z-value from some $z' \neq z$ to z; this process increases $P(z, t + 1)$. The second term corresponds to the sites changing their value from z to some other z-value. During a single time step, a site will change its z-value by an amount equal to the number of times it has been chosen as a (random) neighbor for a relaxing overcritical site. The probability that a site is chosen as a neighbor is equal to $1/N$ in a system of size N. That is, the probability that a site is chosen to be neighbor to k overcritical sites is equal to N^{-k}. In the limit of $N \to \infty$ we can neglect the possibility $k > 1$. In this approximation, a site can receive at most one unit of sand during a time step. The transition probabilities connect z-values that differ by precisely one unit. When a site topples, it induces a change of the z-value of q_c other sites; accordingly, we have

$$W_{z,z';t} = \frac{q_c n_t}{N} \delta_{z-1,z'}, \qquad (5.5)$$

where n_t is the number of topplings in the system at time t and $\delta_{i,j}$ denotes the Kronecker δ-function ($\delta_{i,j} = 0$ when $i \neq j$ and $\delta_{i,j} = 1$ when $i = j$). From (5.4) we obtain

$$P(z, t + 1) - P(z, t) \sim [P(z - 1, t) - P(z, t)]. \qquad (5.6)$$

Thus, in the time-independent stationary state where $P(z, t + 1) = P(z, t) = P(z)$, we conclude that $P(z - 1) = P(z)$ for all $z = 0, 1, 2, \ldots, z_c - 1$. We normalize,

$$\sum_{z=0}^{z_c-1} P(z) = 1, \qquad (5.7)$$

and obtain $P(z) = 1/z_c$ for all $z < z_c$.

From (5.3) we obtain the branching ratio $\sigma = \alpha$. From the expression in (5.1) for the distribution of tree sizes, we see that the nonconservative ($\alpha < 1$) random neighbor BTW model is noncritical. For avalanche sizes $s < s_0$ the avalanche size distribution behaves like a power law, but the exponential cutoff at $s_0(\alpha)$ is *independent* of N. This conclusion agrees with results from simulations of the random neighbor BTW model (Christensen and Olami 1993). In the next section we discuss how the 1992 model of Olami, Feder, and Christensen (OFC) can remain critical even in the nonconservative case.

5.2.2 Earthquake Models

In this section we present a somewhat handwaving mean field calculation of the branching ratio in the nonconservative OFC model defined in Section 4.3. Our aim is to calculate the average number of sites that become overcritical as a result of the energy distributed during the relaxation of an overcritical site (Lise and Jensen 1996). Consider an undercritical site of energy $E < E_c$; the amount αE is distributed to q_c neighbors. Let $P_+(E)$ denote the probability that an undercritical site becomes overcritical as a result of receiving the energy αE. The average number of relaxations induced by the distribution of the q_c packages of energy αE is

$$\sigma = q_c \frac{\int_{E_c}^{\infty} P_+(E) P(E \mid E \geq E_c) \, dE}{\int_{E_c}^{\infty} P(E \mid E \geq E_c) \, dE}. \tag{5.8}$$

The probability that an overcritical site possesses energy E is denoted by $P(E \mid E \geq E_c)$. The probability $P_+(E)$ is equal to the probability that the energy of the receiving site is in the interval $[E_c - \alpha E, E_c]$; this probability is equal to the integral

$$P_+(E) = \int_{E_c - \alpha E}^{E_c} P(\tilde{E}) \, d\tilde{E}, \tag{5.9}$$

where $P(\tilde{E})$ is the probability density of the site variable. Unfortunately, we don't know how to calculate this distribution. As a crude approximation we will simply assume that $P(\tilde{E})$ is uniform on the interval $[0, E_c]$, although we know that this is really not the case (see Figure 5.2).

However, we can hope that this discrepancy is of minor relevance to our argument. In this approximation, $P_+(E) = \alpha E / E_c$. From (5.8) we then obtain

$$\sigma = \frac{q_c \alpha}{E_c} \frac{\int_{E_c}^{\infty} E P(E \mid E \geq E_c) \, dE}{\int_{E_c}^{\infty} P(E \mid E \geq E_c) \, dE} = \frac{q_c \alpha}{E_c} \langle E^+ \rangle, \tag{5.10}$$

where $\langle E^+ \rangle$ is the average energy of the relaxing sites. We can estimate $\langle E^+ \rangle$ in the following way (we put a $+$ superscript on overcritical energy values). The undercritical site j contains the energy $E_j(t)$ at time t. Upon receiving the amount $\alpha E_i^+(t)$ from the overcritical site i, site j becomes overcritical – that is, $E_j^+(t+1) = E_j(t) + \alpha E_i^+(t)$. We average this equation over the possible values of $E_i^+(t)$ and obtain

$$\langle E_j^+(t+1) \rangle = \langle E_j(t) \rangle + \alpha \langle E_i^+(t) \rangle, \tag{5.11}$$

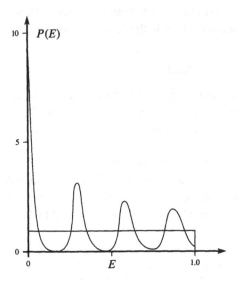

Figure 5.2. Probability density of the site variable in the random neighbor OFC model. (Sketch of data from Lise and Jensen 1996.) The threshold is $E_c = 1$.

where we have assumed that $E_j(t) \in [E_c - \alpha E_i^+(t), E_c]$. Hence, if we again approximate $P(E)$ by a uniform distribution on $[0, E_c]$ then we have

$$\langle E_j(t) \rangle = E_c - \tfrac{1}{2}\alpha \langle E_i^+(t) \rangle. \tag{5.12}$$

The averages cannot depend on the site index (we neglect surface effects) or the instant in time, so $\langle E_j^+(t+1) \rangle = \langle E_i^+(t) \rangle = \langle E^+ \rangle$. From (5.11) we now obtain

$$\langle E^+ \rangle = \frac{E_c}{1 - \alpha/2}. \tag{5.13}$$

The equation for the branching ratio then reduces to

$$\sigma = \frac{q_c \alpha}{1 - \alpha/2}. \tag{5.14}$$

This expression predicts that the branching ratio becomes smaller than unity when

$$\alpha < \alpha_c = \frac{2}{1 + 2q_c}. \tag{5.15}$$

A branching ratio < 1 is equivalent to exponentially decaying avalanche sizes. The prediction by our mean field calculation is, accordingly, that the two-dimensional ($q_c = 4$) OFC model is critical for $\alpha \geq 2/9$. The nearest neighbor OFC model remains critical for values of α smaller than $2/9$. This

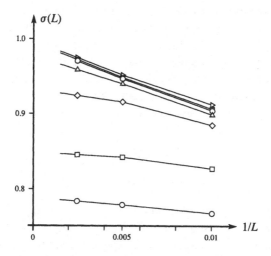

Figure 5.3. Branching ratio in the random neighbor OFC model. (Sketch of data from Lise and Jensen 1996.) The curves represent different levels of conservation. From bottom to top, $\alpha = 0.2$, 0.21, 0.22, 0.225, 0.23, 0.24, and 0.25.

discrepancy is what we expect of mean field predictions. The random neighbor OFC model is expected to be close to the mean field description. In fact, a transition from critical to noncritical behavior is found numerically in the random neighbor model when α becomes smaller than about 2/9 (see Section 4.3.3). The branching ratio of the random neighbor OFC model has also been simulated. The branching ratio $\sigma(L)$ is found for a given system size L^2. The value for infinite system size is obtained by extrapolation. Only when $\alpha \geq 2/9$ does the limit $\lim_{L \to \infty} \sigma(L)$ appear to be equal to unity. A change in behavior for $\alpha \simeq 0.225$ is clearly seen in Figure 5.3. The extent of this agreement is somewhat unexpected, since we used a seemingly inaccurate approximation for the distribution $P(E)$. (See note on p. 124.)

The expression (5.14) leads to a branching ratio *larger* than unity when $\alpha > 2/9$. This implies exponentially growing avalanches – a kind of overcritical behavior. This is not what happens in the simulations of the random neighbor OFC model. In the simulations, $\sigma < 1$ for $\alpha < 2/9$ whereas $\sigma = 1$ in the interval $2/9 < \alpha < 1/4$. It is the additional dissipation at the open boundary that prevents the model from becoming overcritical for $\alpha > 2/9$.

The criterion $\sigma \geq 1$ can be interpreted as a kind of dynamical conservation principle. According to (5.10), $\sigma = 1$ when the average amount of energy $q_c \alpha \langle E^+ \rangle$ fed into the evolving avalanche by the relaxing sites is equal to the lower bound E_c on the loss of energy on the relaxing sites.

Let us try to improve our understanding of why the sandpile model becomes noncritical as soon as a slight amount of dissipation is included, whereas the earthquake model can remain critical even in the presence of dissipation. Consider a slightly generalized sandpile model. Instead of operating with integers as site variables, we assign to every site of our lattice a real number E_i as in the OFC model. The relaxation of the model is defined similarly as in the dissipative version of the BTW model considered in Section 5.2.1. Namely, if $E_i \geq E_c$ then

$$E_i \rightarrow E_i - E_c,$$
$$E_{nb} \rightarrow E_{nb} + \alpha E_c. \qquad (5.16)$$

Here nb denotes the q_c neighbors of the relaxing site i. Energy is lost during the update when $\alpha < 1/q_c$. Notice the difference with respect to the OFC model. In the OFC model, the variable of the relaxing site is set equal to zero, and the amount of energy received by the neighbor sites is a variable quantity that depends on the amount of energy at the relaxing site. In the present model, however, the energy received is always equal to αE_c. This turns out to make a large difference. Let us repeat our mean field calculation of the branching ratio for the model in (5.16). In this case a site with an energy between $E_c - \alpha E_c$ and E_c will topple if it is neighbor to a relaxing site. The branching ratio is therefore given by

$$\sigma = q_c \int_{(1-\alpha)E_c}^{E_c} dE \, P(E) = q_c \alpha. \qquad (5.17)$$

The second equality follows because $P(E) = 1/E_c$ in this model. That $P(E)$ is uniform on $[0, E_c]$ is easily seen by considering the master equation for $P(E)$ in a way similar to what we did in Section 5.2.1. The branching ratio in this model is always less than unity if the model is dissipative, that is, when $\alpha < q_c$. A comparison of (5.17) and (5.10) highlights the importance of the difference between distributing a fixed amount of energy or an amount proportional to the energy of the relaxing site.

There is another difference between the OFC model and the BTW model. The OFC model is driven homogeneously. The BTW model, on the other hand, is driven by adding units of sand at randomly chosen sites. This difference is not responsible for the lack of criticality in the nonconservative BTW model. We do not obtain any kind of OFC-like behavior by driving the model in (5.16) homogeneously. This is seen in simulations (Ghaffari, Lise, and Jensen 1997). Namely, as soon as $\alpha < q_c$, the homogeneously driven model in (5.16) ceases to be critical.

5.2.3 *Diffusive Description of Lattice Gas*

In this section we discuss a somewhat different type of mean field theory. We focus on the equation of motion for the density of particles moving on the lattice in the model defined in Section 4.4. As always in hydrodynamical descriptions, the idea is to imagine the system divided into subvolumes of a size much smaller than any macroscopic scale of the system but large enough to contain a considerable number of particles. In this way we can talk about the density of the particles and how this density depends on time and position.

The fluctuation spectrum observed in the lattice gas can be reproduced by considering diffusion equations for the density of particles (Jensen 1991; Grinstein et al. 1992). The simplest equation of motion one can think of for the density of lattice gas particles is the linear diffusion equation

$$\frac{\partial n(\mathbf{r}, t)}{\partial t} = \gamma \nabla^2 n(\mathbf{r}, t) + \rho(\mathbf{r}, t) \quad \text{for } \mathbf{r} \in \Omega,$$
$$n(\mathbf{r}, t) = \eta(\mathbf{r}, t) \quad \text{for } \mathbf{r} \in S, \tag{5.18}$$

where S is the surface of the region Ω. The power spectrum $S_N(f)$ of

$$N(t) = \int_\Omega d\mathbf{r}\, n(\mathbf{r}, t) \tag{5.19}$$

is easily expressed in terms of the power spectrum of η and ρ by solving (5.18) by use of the Green's function for the equation (Morse and Feshbach 1953; Fiig and Jensen 1993). If the system is driven solely by a delta-correlated *surface* term ($\rho = 0$) the power spectrum behaves as $S(f) \sim 1/f$ for frequencies larger than the inverse diffusion time across the system. On the other hand, if the system contains – in addition to the surface drive – a delta-correlated bulk source ($\rho \neq 0$), then the power spectrum changes behavior to $S(f) \sim 1/f^{3/2}$. This result is not changed by the inclusion of nonlinear terms in (5.18) (see Grinstein et al. 1992). That is, we find the same behavior as in the lattice gas if we assume that the deterministic lattice gas is equivalent to the surface-driven diffusion equation and that the stochastically driven lattice gas corresponds to the bulk-driven case. It is, of course, not easy to derive rigorously from the updating algorithm the form of the noise or driving term needed for a description in terms of effective Langevin or diffusion equations.

The $1/f$ spectrum produced by the surface-driven diffusion equation can probably also be considered as an explanation of the $1/f$ spectrum observed in the surface-driven BTW sandpile model; see Section 4.2.5.

Exercise. The solution to (5.18) is, according to Morse and Feshbach (1953, chap. 7), given by

$$n(\mathbf{r}, t) = \gamma \int dt_0 \int d\vec{S}_0 [G(\mathbf{r}, t \mid \mathbf{r}_0, t) \vec{\nabla} n(\mathbf{r}_0, t_0)$$

$$- n(\mathbf{r}_0, t_0) \vec{\nabla}_0 G(\mathbf{r}, t \mid \mathbf{r}_0, t_0)]$$

$$+ \int dt_0 \int d\mathbf{r}_0 \, G(\mathbf{r}, t \mid \mathbf{r}_0, t_0) \rho(\mathbf{r}_0, t_0), \qquad (5.20)$$

where the first integral is over the surface S and the second over the bulk of the considered domain Ω. The Green's function $G(\mathbf{r}, t \mid \mathbf{r}_0, t_0)$ is the solution to

$$\left[\frac{\partial}{\partial t} - \gamma \vec{\nabla}^2 \right] G(\mathbf{r}, t \mid \mathbf{r}_0, t_0) = \delta(\mathbf{r} - \mathbf{r}_0) \delta(t - t_0) \qquad (5.21)$$

in the domain Ω. Assume the homogeneous Dirichlet condition

$$G(\mathbf{r}, t \mid \mathbf{r}_0, t_0) = 0 \quad \text{for } \mathbf{r}_0 \in S. \qquad (5.22)$$

Extract the frequency dependence of $S_N(f)$ in the following three cases:

(a) delta-correlated surface drive,

$$\langle \eta(\mathbf{r}, t) \eta(\mathbf{r}', t') \rangle \sim \delta(\mathbf{r} - \mathbf{r}') \delta(t - t') \quad \text{and} \quad \rho(\mathbf{r}, t) = 0; \qquad (5.23)$$

(b) delta-correlated bulk drive,

$$\langle \rho(\mathbf{r}, t) \rho(\mathbf{r}', t') \rangle \sim \delta(\mathbf{r} - \mathbf{r}') \delta(t - t') \quad \text{and} \quad \eta(\mathbf{r}, t) = 0; \qquad (5.24)$$

(c) surface and bulk drive, both delta-correlated, acting simultaneously.

5.2.4 Forest Fire Model

A mean field description of the forest fire model defined in Section 4.5 was constructed by Christensen et al. (1993). The theory consists of a set of equations of motion for the three densities characterizing the state of the model: the density of trees $\rho_t(t)$, the density of trees in fire $\rho_f(t)$, and finally the density of empty sites $\rho_e(t)$. Assume that there are many more green trees than there are trees in fire. In this limit, a tree can catch fire by interacting with a single burning site or by being hit by lightning; the latter happens at a rate f. The time evolution of the three densities is given by the following set of equations:

$$\rho_e(t + 1) - \rho_e(t) = -p\rho_e(t) + \rho_f(t) \qquad (5.25)$$

$$\rho_t(t + 1) - \rho_t(t) = -(f + q_c \rho_f(t))\rho_t(t) + p\rho_e(t) \qquad (5.26)$$

$$\rho_f(t + 1) - \rho_f(t) = -\rho_f(t) + (f + q_c \rho_f(t))\rho_t(t). \qquad (5.27)$$

The change in the number of empty sites (5.25) consists of two terms. The first describes the loss of empty sites by the growing of new green trees at a rate

p times the number of available empty sites. The second term denotes the addition of new empty sites created by the burning down of trees. The change in the number of trees (5.26) consists of a depletion of tree sites by spontaneous lightning and the possible spreading of fires from one† of the $q_c\rho_f$ neighbor sites in fire. New trees are added at the rate p times the number of available empty sites. Finally, (5.27) represents the change in the number of sites in fire. The sites in fire at time t have become empty sites at time $t+1$, which explains the first term on the right-hand side of (5.26). Green trees add to the number of sites in fire at a rate f times the number of green trees able to be hit by lightning; this is the second term of the equation. Finally, the term entering (5.26) as a depletion term – namely, the number of green trees catching fire from a burning neighbor site – enters in (5.27) as a positive contribution. Because all sites of the model fall into one of three complementary categories *tree, fire,* or *empty,* the densities must add up to unity. That is,

$$\rho_t(t) + \rho_f(t) + \rho_e(t) = 1. \tag{5.28}$$

We can use this constraint to eliminate one of the variables in the dynamical equations (5.25)–(5.27). Furthermore, we will consider time as a continuous variable and replace the time differences with derivatives. We then arrive at the following set of dynamical equations:

$$\begin{aligned}
\dot{\rho}_t &= -f\rho_t - q_c\rho_t\rho_f + p(1 - \rho_t - \rho_f) \equiv F(\rho_t, \rho_f), \\
\dot{\rho}_f &= -\rho_f + f\rho_t + q_c\rho_t\rho_f \equiv G(\rho_t, \rho_f).
\end{aligned} \tag{5.29}$$

We can apply standard linear stability analysis to this set of equations (Boyce and Di Prima 1965). First we identify the fixed point of the equation by solving for $\dot{\rho}_t = 0$ and $\dot{\rho}_f = 0$. There is only one solution compatible with the requirement that $\rho_t \leq 1$:

$$\begin{aligned}
\rho_t^* &= \frac{1}{2q_c}(1 + q_c + \kappa - [\kappa^2 + 2(1 + q_c)\kappa + (q_c - 1)^2]^{1/2}), \\
\rho_r^* &= \frac{p}{1 + p}(1 - \rho_t),
\end{aligned} \tag{5.30}$$

where $\kappa \equiv f(1 + p)/p$. The behavior in the vicinity of the fixed point (ρ_t, ρ_f) is determined by the eigenvalues λ of the determinant equation

$$\begin{vmatrix} \partial F/\partial \rho_t - \lambda & \partial F/\partial \rho_f \\ \partial G/\partial \rho_t & \partial G/\partial \rho_f - \lambda \end{vmatrix} = 0. \tag{5.31}$$

This equation is of the form $\lambda^2 + B\lambda + C = 0$ with

---

† Remember that we consider the limit of very few fires compared to trees, which is why we neglect the possibility of more than one burning neighbor site.

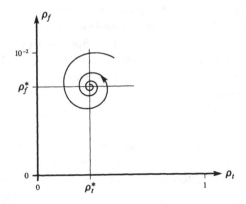

Figure 5.4. Phase plane trajectories of the mean field forest fire model.

$$B = 1 + f + p + q_c(\rho_f^* - \rho_f^*),$$
$$C = f(1 + p) + p(1 + q_c(\rho_f^* - \rho_t^*)) + q_c\rho_f^*.$$

In the double limit $f/p \to 0$ and $p \to 0$, the two roots of this parabola are complex with negative real parts. We conclude that the fixed point of (5.30) is an attracting spiral. The phase plane trajectories are sketched in Figure 5.4. The prediction for density of trees in the asymptotic regime is $\rho_t^* = 1/q_c$ in the limit $f = 0$.

The forest fire can also be considered as a branching process in the same way that the sandpile model was in Section 5.2.1. The branching ratio σ is equal to the average number of trees ignited by a fire spreading from one burning tree. A tree can ignite as few as none or as many as q_c of its neighbors. The calculation of σ is given by the same expression as in (5.2) and (5.3); that is,

$$\sigma = \sum_{n=0}^{q_c} nK_{q_c,n}\rho_t^n(1 - \rho_t)^{1-n}$$
$$= 1 - \frac{\kappa}{q_c - 1} + O(\kappa^2). \tag{5.32}$$

In other words, the forest fire becomes a critical branching process in the limit $f \to 0$. The distribution of sizes of fires is given by (5.1), so the value for the exponent of the size distribution is $\tau = 3/2$.

The forest fire can also be considered as a percolation process (Christensen et al. 1993). This description is still "mean field" in the sense that spatial

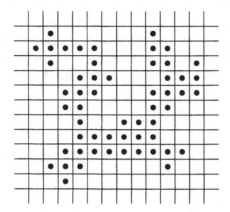

Figure 5.5. A connected cluster of trees.

correlations are neglected and trees are considered as uniformly randomly distributed over the lattice at a density equal to ρ_t. In this approximation, the sizes of fires are simply given by the expression in percolation theory for the sizes of connected clusters (Stauffer and Aharony 1992). Consider a lattice in d dimensions. Choose a site at random and plant a tree at the site. Continue this process until the fraction ρ_t of all sites contains a tree. A *cluster* is defined as a set of trees that are all connected through nearest neighbor links (see Figure 5.5). If one tree in the cluster catches fire then the whole cluster will eventually burn down, because the fire will be able to spread through nearest neighbor links to all trees in the cluster. This way of considering the forest fire model allows one to make use of several results in percolation theory – in particular, the result that the upper critical dimension is 6. This means that the values for power-law exponents derived in mean field theory should be correct in dimensions $\geq$ 6. In fact, simulations in dimensions 1–6 (Christensen et al. 1993) and in dimensions 1–8 (Clar et al. 1994) confirm this expectation.

5.2.5 Model of Biological Evolution

In this section we present a mean field analysis of the Bak–Sneppen evolution model (Bak and Sneppen 1993). Our discussion is a simple generalization of the mean field theory by Flyvbjerg, Sneppen, and Bak (1993). In the evolution model described in Section 4.6.2, only the smallest barrier undergoes a mutation. We consider a model in which $M \geq 1$ smallest barriers are simultaneously updated.

Our mean field description assumes a random neighbor version of the model and, in addition, neglects correlations between barriers.† The analysis then reduces to combinatorics. Let $p(B)\,dB$ denote the probability that an arbitrarily chosen site is found to have a barrier variable in the interval $[B, B + dB]$. The probability that a barrier B is among the M smallest barriers is given by

$$p_{\min}(B) = C^{-1} \sum_{i=1}^{M} p_i(B),$$ (5.33)

where

$$p_i(B) = \frac{L!}{(L - i)!\,(i - 1)!} P^{i-1}(B) p(B) Q^{L-i}(B),$$

$$P(B) = \int_0^B ds\, p(s),$$ (5.34)

$$Q(B) = 1 - P(B),$$

and C is a normalization constant given by

$$C = \sum_{q=1}^{M} \sum_{n=0}^{q-1} q K_{L,q} K_{q-1,n}(-1)^n (L - q + n + 1)^{-1}$$ (5.35)

in terms of the binomial coefficients $K_{p,q}$. The function $p_i(B)$ is the probability that a barrier B is precisely the ith smallest barrier. The following argument makes that clear.

We can calculate $p_i(B)$ by imagining that we consecutively visit the L sites of our one-dimensional string, making note of the B-values of the sites. We must ascertain whether a given B_0 is precisely the ith smallest of all B-values. We can approach this as if we first make $i - 1$ draws and obtain B-values that are smaller than B_0. The probability that this happens is given by

$$p_1 = \frac{L(L - 1) \cdots (L - (i - 2))}{(i - 1)!} (P(B_0))^{i-1}.$$ (5.36)

The ith time we make a draw there are $L - (i - 2)$ sites that we have not yet visited. The probability that the outcome of the ith draw equals B_0 is

$$p_2 = (L - (i - 1)) p(B_0).$$ (5.37)

The remaining $L - i$ times we draw, we must obtain B-values that are greater than B_0. The probability for this is

$$p_3 = \frac{(L - i)(L - i - 2) \cdots 1}{(L - i)!} Q^{L-i}.$$ (5.38)

The expression in (5.34) follows since $p_i(B_0) = p_1 p_2 p_3$.

---

† The random neighbor version of the Bak–Sneppen evolution model can be solved exactly; see de Boer et al. (1994).

The model is updated in the following way. First we draw the M smallest barriers from the distribution $p_{min}(B)$. Next, $K - M$ sites are chosen at random (random neighbor model). It seems natural to put $K = 3M$ in order to describe the model discussed in Section 4.6.2; however, for the moment we will keep K general. The evolution equation for $p(B, t)$ is given by

$$p(B, t + 1) = p(B, t) - mp_{min}(B, t)$$
$$- \frac{K - M}{L - M}[p(B, t) - mp_{min}(B, t)] + \frac{K}{L}, \quad (5.39)$$

where $m = M/L$. It is perhaps easier to see the meaning of this equation if one multiplies by L:

$$Lp(B, t + 1) = Lp(B, t) - Mp_{min}(B, t)$$
$$- \frac{K - M}{L - M}[Lp(B, t) - Mp_{min}(B, t)] + K. \quad (5.40)$$

Now $Lp(B, t + 1)$ is the number of sites with B-values equal to B at time $t + 1$. This number is calculated as being equal to the number of sites at time t with this B-value, corrected for the number of sites that have changed their B-value owing to the action of the update. The second term on the right-hand side describes the probability that a barrier value B is among the M smallest and therefore subject to update. Among the $L - M$ remaining sites, $K - M$ sites are chosen at random and updated. In each of these $K - M$ "draws," the probability that a chosen site has a barrier value of B is equal to $(Lp(B, t) - Mp_{min}(B, t))/(L - M)$. This is because there are $Lp(B, t)$ sites of barrier B on the system, but a barrier is removed from the system if it is among the minimal barriers – hence the term $Mp_{min}(B, t)$. The last term on the right-hand side of (5.40) arises because each of the K draws can potentially produce a barrier equal to B in time step $t + 1$. The probability of this happening is equal to $KP_0(B)$, where $P_0(B)$ is the probability density with which new barriers are generated. Since we assume a uniform distribution on the interval $[0, 1]$, we have $P_0(B) = 1$.

The stationary solution to this equation is readily solved by the same procedure as used by Flyvbjerg et al. (1993) in the $M = 1$ case. For the stationary solution, $p(B, t + 1) = p(B, t)$ and therefore (5.39) can be written as

$$p(B) = \frac{k - 1}{k - m}p_{min}(B) + \frac{1 - m}{k - m}, \quad (5.41)$$

where $k = K/L$. Use that $p(B) = -dQ/dB$ and express $p_{min}(B)$ in terms of (5.33) and (5.34). Multiplying through by dB and integrating, one finally obtains

$$Q = \frac{1-m}{1-m/k}(1-B) + \frac{k-1}{k/m-1}\frac{1}{C}$$

$$\times \sum_{q=1}^{M} \sum_{n=1}^{q-1} \frac{L!}{(L-q)!\,(q-n-1)!\,n!} \frac{Q^{L-q+n+1}}{L-q+n+1}. \tag{5.42}$$

In the limit of large system size L and small m, the solution $p(B)$ to this equation has the form

$$p(B) = \begin{cases} p_1 & \text{if } 0 < B < B_c, \\ p_2 & \text{if } B_c < B < 1, \end{cases} \tag{5.43}$$

where $p_1 \sim 1/L$ vanishes in the limit of large systems. This size dependence of p_1 is identical to the one found in simulations of the nearest neighbor version of the model (Datta et al. 1997). In the limit $L \to \infty$, the plateau value above B_c is given by $p_2 = (1-m)/(1-m/k) + o(1/L)$ and the threshold value by $B_c = (m/k - m)/(1-m) + o(1/L)$.

We notice that the mean field theory reproduces the steplike shape of the distribution function in simulations (see Figure 4.17). The mean field theory predicts the vanishing of the lower plateau value p_1 with increasing system size. The decrease in p_2 with increasing M is seen in simulations. The mean field theory predicts a smooth decrease in B_c with increasing M. Numerically, one finds a precipitous decrease in B_c for small M-values as M is increased from $M = 1$ (Datta et al. 1997). This discrepancy is due to spatial correlations between the updated sites. The mean field expression for B_c predicts that the behavior of $B_c(M)$ in the limit of $L \to \infty$ is determined by the ratio M/K. Let $K(M)$ denote the average of the actual number of sites with *different* positions involved in the update of the M sites with the M smallest B-values. Neglecting spatial correlations, one has $K(M) = 3M$ leading to a B_c that is independent of M. One can understand the drop in B_c as being caused by an attraction between minimal sites. That minimal sites tend to cluster is of course seen from the distribution of distances between consecutive updates, as discussed for the case $M = 1$ in Section 4.6.2. The attraction arises because the B-value of a newly updated site is distributed uniformly on the interval $[0, 1]$. The site variable of a randomly chosen site is distributed according to $p(B)$, which is vanishingly small for $B < B_c$. Thus, a newly updated site is more likely to possess a small B-value. This attraction leads to a lower effective value of the ratio M/K and thereby to a drop in B_c.

5.3 Exact Solution of the Abelian Sandpile

We now leave the somewhat fuzzy world of mean field thinking and enter the sharp, rigorous realm of exact solutions.

The exact solution method developed by Dhar and collaborators (Dhar and Ramaswamy 1989; Dhar 1990; Dhar and Majumdar 1990) involves driving the sandpile models in terms of operators acting on a configuration space. The one crucial assumption needed in order to make the algebra of the operators tractable is that the algorithm of the model be Abelian. This means that the *order* in which the perturbations to the model are applied is irrelevant. It should not matter whether we first add sand to site A followed by an addition of sand to site B or reverse the order. The Abelian character is obtained if the update depends only on the state of the site to be updated and so is independent of the state of the surrounding sites. Moreover, the action of the update must be independent of the actual value of the dynamical field at the updated site.

5.3.1 The Δ Matrix and the Probability Measure on the Configuration Space

The following language is useful for purposes of highlighting the nature of the algebra behind the algorithm of the sandpile models. Consider a system consisting of N sites $1, 2, 3, \ldots, N$. To each site is assigned an integer variable z_i. The system is driven according to two rules as follows.

(1) *Addition Rule.* To add a particle, select a site (say, i) at random and let

$$z_i \rightarrow z_i + 1. \tag{5.44}$$

Define N threshold values z_{ic} and an $N \times N$ matrix Δ. The elements of the matrix are assumed to fulfill the following conditions:

$$
\begin{aligned}
\text{(a)} \quad & \Delta_{ii} > 0 \quad \forall i; \\
\text{(b)} \quad & \Delta_{ij} \leq 0 \quad \forall i \neq j; \\
\text{(c)} \quad & \sum_{j=1}^{N} \Delta_{ij} \geq 0 \quad \forall i.
\end{aligned}
\tag{5.45}
$$

Overcritical sites are updated according to the toppling rule.

(2) *Toppling Rule.* If $z_i > z_{ic}$, then $z_j \rightarrow z_j - \Delta_{ij}$ for $j = 1, 2, \ldots, N$.

Think of z_i as the number of particles present at position i. Condition (a) of the Δ-matrix ensures that an overcritical site always lowers its number of resident particles as a result of the update. Accordingly, condition (b) ensures that the toppling of the overcritical site leads to particles being added to the surrounding sites. Hence, we can think of the update as particles being moved from site i

to certain neighboring sites. The neighbors assigned to a site i are, of course, given by the values of j for which Δ_{ij} is different from zero. Finally, condition (c) secures that no particles are created during the update. However, particles *can* disappear from the system (say, over the edge); this happens whenever $\sum_{j=1}^{N} \Delta_{ij} > 0$. Obviously, particles must disappear from the system in order for a stationary state to exist.

Exercises. (1) Write down the matrix Δ for the two-dimensional BTW sandpile model as defined in Section 4.2.

(2) To what extent are the results derived in this section relevant to the BTW sandpile?

The set $\mathcal{S}$ of stable configurations is defined as

$$\mathcal{S} = \{\, C = \{z_i\} \mid 1 \le z_i \le z_{ic} \; \forall i \,\}.$$

For convenience we assume $z_{ic} = \Delta_{ii}$. Next we define operators $a_i : \mathcal{S} \to \mathcal{S}$. The operator a_i adds one unit to site i according to the addition rule (1). If necessary, the resulting configuration is then relaxed by applying rule (2) until all sites once again assume z-values below $z_{ic} = \Delta_{ii}$. The operators a_i commute on $\mathcal{S}$. To see this, take a configuration $C_0 = \{z_i^0\}$ in which two sites $i = a, b$ are overcritical: $z_a > \Delta_{aa}$ and $z_b > \Delta_{bb}$. We topple these two sites (and only these two sites) according to rule (2). The effect is that the configuration $C_0 \to C_1 = \{z_i^1\}$, where the z-values of the resulting configurations are given by $z_i^1 = z_i^0 + \Delta_{ai} + \Delta_{bi}$. This result does not depend on which site, a or b, we imagine was toppled first. Because an avalanche consists simply of successive applications of rule (2), we see that the configuration produced by the operation of $a_i a_j$ on any configuration $C \in \mathcal{S}$ is the same as the one produced by $a_j a_i$. We conclude that $[a_i, a_j] \equiv a_i a_j - a_j a_i = 0$.

Among all configurations, those that are relevant to the behavior of the system in the stationary state are the recurrent configurations. The set of recurrent configurations $\mathcal{R} \subseteq \mathcal{S}$ is defined as the set of configurations for which we always (eventually) return to the same configuration after repetitive application of the operators a_i. More precisely,

$$\mathcal{R} = \{\, C \in \mathcal{S} \mid \exists m_i : a_i^{m_i} C = C \; \forall i \,\}. \tag{5.46}$$

Here m_i are integers. Note that $a_i : \mathcal{R} \to \mathcal{R}$. Namely, let $C \in \mathcal{R}$; then $C = a_i^{m_i} C$ and, accordingly, $a_i^{m_i}(a_{i_0} C) = a_{i_0} a_i^{m_i} C = a_{i_0} C$. In other words, $a_{i_0} C$ is recurrent. All configurations can be divided into transient or recurrent configurations. Only the recurrent configurations occur with probability other than zero in the stationary state of the system. The operator algebra defined here allows a calculation of the probability measure on the set of recurrent configurations. We now describe how this is achieved.

When restricted to the set $\mathcal{R}$, all operators a_i can be inverted. This follows immediately from the recurrence property, which may be stated as follows: for all $i = 1, 2, \ldots, N$,

$$a_i^{-1}C = a_i^{m_i-1}C. \tag{5.47}$$

Since a_i can be inverted on $\mathcal{R}$, there is a unique correspondence between the image $a_i C$ and the configuration C. This means that, assuming $a_i C_1 = a_i C_2$ for two configurations in $\mathcal{R}$, it follows that $C_1 = C_2$. From this property it can be shown that all configurations in $\mathcal{R}$ occur with equal probability in the stationary state. Let us sketch the proof.

Consider the evolution of the system. Let $P(C, t)$ denote the probability that at time t the system is found in the configuration C. Let $W(C \to C')$ denote the transition probability that configuration C changes into configuration C'. The master equation (Reif 1965) for the evolution of the probability measure $P(C, t)$ is

$$P(C, t+1) = P(C, t) - \sum_{C' \in \mathcal{R}} P(C, t) W(C \to C')$$
$$+ \sum_{C' \in \mathcal{R}} P(C', t) W(C' \to C). \tag{5.48}$$

The transitions in the system are induced by the (random) application of the operators a_i. The transition probabilities can be expressed as

$$W(C \to C') = \sum_{i=1}^{N} P(a_i) P(a_i C = C'), \tag{5.49}$$

where $P(a_i) = p_i$ is the probability that the perturbation at time t is applied to site number i. The factor $P(a_i C = C')$ denotes the probability that the configuration C' is hit when a_i operates on the configuration C. Since a_i is invertible on $\mathcal{R}$, we simply have

$$P(a_i C = C') = \begin{cases} 1 & \text{if } C = a_i^{-1} C', \\ 0 & \text{otherwise.} \end{cases} \tag{5.50}$$

We substitute (5.49) and (5.50) into the master equation (5.48) and obtain, after performing the sum over C',

$$P(C, t+1) = P(C, t) - \sum_{i=1}^{N} p_i \{ P(a_i C, t) - P(a_i^{-1} C, t) \}. \tag{5.51}$$

From this equation it is clear that a time-independent uniform distribution, $P(C, t) = \text{const.}$ for all $C \in \mathcal{R}$, is a stationary solution to the master equation.

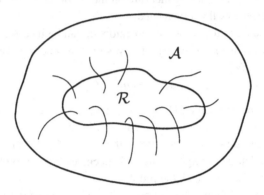

Figure 5.6. The set $\mathcal{A}$ of all configurations that can be reached from the configuration $\mathcal{R}$ by adding particles.

The master equation (5.48) for Markovian evolution has a unique invariant, or time-independent, state. We want to identify the SOC state of the system as the stationary state. Hence, we can conclude that the SOC state of the model consists of the state in which the evolution of the system

$$\cdots \to C_{t-1} \to C_t \to C_{t+1} \to \cdots \tag{5.52}$$

occurs between states in $\mathcal{R}$ and where all $C \in \mathcal{R}$ are visited with the same probability – that is, $P(C, t) = 1/|\mathcal{R}|$. To determine the probability measure we need to calculate $|\mathcal{R}|$, the number of elements in the set $\mathcal{R}$.

Dhar (1990) has constructed a clever and illuminating argument that allows us to calculate $|\mathcal{R}|$. The trick is to consider $\mathcal{R}$ as a subset of a larger set $\mathcal{A}$. The set $\mathcal{A}$ is defined as all those configurations that can be reached by adding particles to configurations in $\mathcal{R}$ (see Figure 5.6). Obviously, adding particles to configurations in $\mathcal{R}$ may produce overcritical sites, but for now we perform no relaxation. An equivalence relation can be defined on $\mathcal{A}$. Namely, two configurations $C \in \mathcal{A}$ and $C' \in \mathcal{A}$ are said to be *equivalent* $(C \equiv C')$ if, when toppled, they relax to the same stable configuration in $\mathcal{R}$.

We now define toppling operators T_n. The effect of T_n on a configuration $C = \mathbf{z} = \{z_i\}$ is to perform the toppling rule (2) if site n is overcritical:

$$T_n\{z_i\} = \begin{cases} \{z_i - \Delta_{ni}\} & \text{if } z_n > \Delta_{nn}, \\ \{z_i\} & \text{otherwise.} \end{cases} \tag{5.53}$$

We can now easily show that $C \equiv C'$ if and only if there exist sequences $T_{n_1} T_{n_2} \cdots T_{n_m}$ and $T_{n'_1} T_{n'_2} \cdots T_{n'_{m'}}$ such that

$$T_{n_1} T_{n_2} \cdots T_{n_m} \mathbf{z} = T_{n'_1} T_{n'_2} \cdots T_{n'_{m'}} \mathbf{z}'. \tag{5.54}$$

By successive application of (5.53) we see that (5.54) corresponds to

$$z_j - \sum_{i=1}^{m} \Delta_{a_i j} = z' - \sum_{i=1}^{m'} \Delta_{a'_i j} \tag{5.55}$$

$$\Downarrow$$

$$z'_j = z_j - \sum_{i=1}^{N} r_i \Delta_{ij}, \tag{5.56}$$

where r_i are integers determined by the number of times T_i occurs in the products in (5.54).

Let $\hat{\mathbf{e}}_i$, $i = 1, \ldots, N$ denote the unit vectors in $\mathbb{Z}^N$ (here $\mathbb{Z}$ denotes the set of integers). Let $\mathbf{n}_i = \sum_{j=1}^{N} \Delta_{ij} \hat{\mathbf{e}}_j$. Define, as usual,

$$\mathrm{span}\{\mathbf{n}_1, \mathbf{n}_2, \ldots, \mathbf{n}_N\} = \left\{ \mathbf{v} \in \mathbb{Z}^N \ \middle| \ \mathbf{v} = \sum_{i=1}^{N} \alpha_i \mathbf{n}_i, \ \alpha_i \in \mathbb{Z} \right\}. \tag{5.57}$$

By use of these definitions and (5.56), we conclude that

$$C \equiv C' \iff \mathbf{z} - \mathbf{z}' \in \mathrm{span}\{\mathbf{n}_1, \ldots, \mathbf{n}_N\}. \tag{5.58}$$

This equation tells us that we obtain all elements of an equivalence class C by picking one element $\mathbf{z} \in C$ and then adding successively all the vectors in $\mathrm{span}\{\mathbf{n}_1, \ldots, \mathbf{n}_N\}$ to $\mathbf{z}$. Hence, the number of different equivalence classes is equal to the number of different vectors $\mathbf{z}$ within the unit cell of the space $\mathrm{span}\{\mathbf{n}_1, \ldots, \mathbf{n}_N\}$. By construction, each equivalence class contains precisely one element from $\mathcal{R}$. Hence, $|\mathcal{R}|$ is equal to the number of elements in the unit cell of $\mathrm{span}\{\mathbf{n}_1, \ldots, \mathbf{n}_N\}$, that is, the unit cell's volume. This volume is simply equal to $\det \Delta$, since the unit cell of $\mathrm{span}\{\mathbf{n}_1, \ldots, \mathbf{n}_N\}$ is obtained as the image of the unit cell of $\mathrm{span}\{\hat{\mathbf{e}}_1, \ldots, \hat{\mathbf{e}}_N\}$ under the linear mapping $\Delta \colon \mathbb{Z}^N \to \mathbb{Z}^N$ defined in terms of the Δ-matrix. Hence, we have shown that

$$P(C, t) = \frac{1}{|\mathcal{R}|} = \frac{1}{\det \Delta}. \tag{5.59}$$

5.3.2 Correlation Functions

The correlations in the sandpile model are also determined by the toppling matrix Δ. We define the two-point correlation function

$$
\begin{aligned}
G_{ij} = \ &\text{the average number of topplings at site } j \\
&\text{induced by avalanches released when a} \\
&\text{particle is added at site } i.
\end{aligned}
\tag{5.60}
$$

The relation between G and Δ can be established from the balance between $J_{\text{in}}(j)$, the average flux of particles into a site j, and $J_{\text{out}}(j)$, the average flux of particles out of the same site. From the toppling rule (2) we can express the flow induced by the addition of a particle at site i by the following expressions:

$$J_{\text{in}}(j) = \sum_{k \neq j} G_{ik}(-\Delta_{kj}) + \delta_{ij}, \qquad (5.61)$$

$$J_{\text{out}}(j) = G_{ij}\Delta_{jj}. \qquad (5.62)$$

We equate the two fluxes and obtain

$$\sum_{k=1}^{N} G_{ik}\Delta_{kj} = \delta_{ij}, \qquad (5.63)$$

from which we conclude that

$$G_{ij} = [\Delta^{-1}]_{ij}. \qquad (5.64)$$

This expression for G_{ij} can, in particular, be used to determine the asymptotic behavior of toppling correlations for the nearest neighbor BTW model on a d-dimensional cubic lattice. We neglect the boundary of the system. For bulk sites i and j, the matrix Δ is given by

$$\Delta_{ij} = \begin{cases} 2d & \text{if } i = j, \\ -1 & \text{if } i \text{ and } j \text{ denote nearest neighbor sites}, \\ 0 & \text{otherwise.} \end{cases} \qquad (5.65)$$

We will make use of the continuum approximation in order to obtain the asymptotic behavior of Δ_{ij}^{-1}. It is easy to see that Δ, defined in (5.65), can be considered as the discrete version of the Laplacian operator in d dimensions. For instance, consider (with $d = 1$ for ease of notation) the equation

$$\nabla^2 f(x) = \delta(x). \qquad (5.66)$$

Confine x to a set of discrete values $x = na$, where n runs through all integers and a is a constant. The discrete version of (5.66) is obtained by making the following substitutions:

$$\delta(x) \to \delta_{n,0},$$
$$\partial_x f(x) \to [f((n+1)a) - f(na)]/a,$$
$$\Downarrow$$

$$\partial^2 f(x) \to [f((n+1)a) - f(na)]/a - [f(na) - f((n-1)a)]/a$$
$$= [f((n+1)a) + f((n-1)a) - 2f(na)]/a^2.$$

Compare the last expression to (5.65). We can consider G_{ij} as the inverse of the Laplace operator or, in other words, as the solution to Poisson's equation in d dimensions:

$$\nabla^2 G(r) = \delta^d(r). \tag{5.67}$$

We make use of the fact that G_{ij} in the bulk will depend only on the relative distance r between the positions of site i and site j. The behavior of the solution to (5.67) is well known and readily obtained by Fourier transformation. Let

$$\hat{G}(k) = \int_{-\infty}^{\infty} d^d\mathbf{r}\, G(r) e^{i\mathbf{k}\cdot\mathbf{r}}. \tag{5.68}$$

From (5.67) we have $\hat{G}(k) = 1/k^2$. We Fourier transform back again and obtain

$$G(r) = \int_{-\infty}^{\infty} \frac{d^d\mathbf{k}}{[2\pi]^d} k^{-2} e^{-i\mathbf{k}\cdot\mathbf{r}}. \tag{5.69}$$

We make the substitution $\mathbf{u} = r\mathbf{k}$ and see that $G(r) \sim r^{2-d}$ – that is, algebraic correlations as expected for a system in a critical state.

The method developed by Dhar is very powerful and has been used by several groups to obtain a long list of interesting results. The height–height correlations $\langle z_i z_j \rangle - \langle z_i \rangle^2$ are found to decay as r^{-2d} (Majumdar and Dhar 1991). Dhar (1990) calculated the average number of topplings per added particle for the nearest neighbor Abelian model on a square lattice of size $L \times L$ with open boundaries: $\langle T \rangle \sim L^2$. This method has also been used to calculate distribution functions for models defined on a Bethe lattice (Dhar and Majumdar 1990). The probability that an avalanche consists of n topplings scales like $n^{-3/2}$ and the probability density for the avalanche lifetimes scales like T^{-2}. Priezzhev, Ktitarev, and Ivashkevich (1996) have calculated the probability density for the avalanche sizes $P(s) \sim s^{-\tau}$, where s is the number of distinct sites toppled during the avalanche. These authors make use of the formalism described here, the notion of inverse avalanches (Dhar and Manna 1994), and the relation to spanning trees. The result is $\tau = 5/4$. This value is in excellent agreement with the result $\tau = 1.253$ obtained by Pietronero, Vespignani, and Zapperi (1994) from the dynamically driven renormalization group calculation (described in Section 5.5). We should also mention that the Δ-matrix formalism has been generalized by Gabrielov to the case where the dynamical field z_i is allowed to assume real values (and not only integer values). This generalization is relevant to models of crack formation and earthquakes (Gabrielov 1993; Gabrielov, Newman, and Knopoff 1994).

The Search for a Formalism

5.4 Langevin Equations

5.4.1 Conservative Models

The BTW sandpile cellular automaton can be considered as a discrete version of a diffusion process (Wiesenfeld et al. 1989). The easiest way to see this is to write down the updating algorithm of the model in the form of an equation of motion. Let us here consider a slightly generalized version of the BTW model, namely the Zhang model (Zhang 1989). The model is driven by adding a random increment $\eta \in [0, 1]$ to the dynamical variable $E(\mathbf{r}, t)$ at a randomly chosen lattice position $\mathbf{r}$. The variable E is taken to be continuous (i.e., to be a real number). When $E(\mathbf{r}, t)$ becomes larger than the threshold value E_c, the site variables are updated as follows:

$$
\begin{aligned}
E(\mathbf{r}, t+1) &\to 0, \\
E(\mathbf{r}_{nn}, t+1) &\to E(\mathbf{r}_{nn}, t) + E(\mathbf{r}, t)/q_c,
\end{aligned}
\tag{5.70}
$$

where $\mathbf{r}_{nn}$ denotes the $q_c = 2d$ nearest neighbor sites on a d-dimensional cubic lattice. Time is discrete. The updating can be summarized in the following equation of motion:

$$
\begin{aligned}
E(\mathbf{r}, t+1) = \ &E(\mathbf{r}, t)[1 - \Theta(E(\mathbf{r}, t) - E_c)] \\
&+ \sum_{\mathbf{r}_{nn}} \frac{1}{q_c} E(\mathbf{r}_{nn}, t) \Theta(E(\mathbf{r}_{nn}, t) - E_c) + \eta(\mathbf{r}, t).
\end{aligned}
\tag{5.71}
$$

A word of caution is in order about the driving term $\eta(\mathbf{r}, t)$. The SOC sandpile models are often driven in the following way. The random addition of extra "energy" takes place only in between avalanche activity. When a site becomes overcritical and an avalanche is released, the external addition of energy is switched off. In other words, the term $\eta(\mathbf{r}, t)$ is taken to be *zero* during the evolution of the avalanches. The argument for the relevance of this type of driving is the separation of time scales. Think of avalanches of snow in the Alps. The snow slopes are built up gradually during the weeks and months of winter, or even over years, whereas the duration of the avalanches themselves is measured in seconds. In this case we choose to neglect any snow that might be falling during the avalanches.

The continuum limit of the discrete equation of motion in (5.71) is easy to derive:

$$
\frac{\partial E(\mathbf{r}, t)}{\partial t} = \tilde{D} \nabla^2 [E(\mathbf{r}, t) \Theta(E(\mathbf{r}, t) - E_c)] + \eta(\mathbf{r}, t),
\tag{5.72}
$$

where the diffusion constant is given formally, in terms of the lattice spacing a_0, the time step Δ, and the coordination number q_c, as $\tilde{D} = a_0^2/(\Delta q_c)$. This expression for $\tilde{D}$ should not be taken too literally. One should rather consider (5.72) as the result of a procedure of coarse graining in which the microscopic site variable of (5.71) has been replaced by a new variable that is equal to the local average of the microscopic variable. Thus we have performed the following substitution in going from (5.71) to (5.72):

$$E(\mathbf{r}, t) \rightarrow \bar{E}(\mathbf{r}, t) \equiv \frac{1}{n} \sum_{i=1}^{n} E(\mathbf{r}_i, t), \qquad (5.73)$$

where the sum is over a small neighborhood of the position $\mathbf{r}$. This procedure is not easy to perform rigorously. However, an equation of the form (5.72) is heuristically expected to be the result of the procedure. The exact relation to the microscopic parameters is lost, and the properties of the noise term in (5.72) will have to be obtained from general arguments.

The coarse-grained equation of motion formally resembles a Langevin equation or a stochastically driven differential equation. Standard renormalization group (RG) procedures have been developed to calculate the asymptotic properties of solutions of such equations (Forster, Nelson, and Stephen 1977; Medina et al. 1989). One feature makes the application of the technique difficult – namely, that the Heaviside step function

$$\Theta(x) = \begin{cases} 1 & \text{if } x \geq 0, \\ 0 & \text{if } x < 0, \end{cases} \qquad (5.74)$$

is a strongly singular function. The RG technique can be applied when the equation's right-hand side contains terms that are analytic functions of E and derivatives of E. Hence, in order to apply the RG method, we need an analytic representation of the Θ-function (see Diaz-Guilera 1994). The trick is to find an analytic function $f(x)$ for which both limits $\lim_{x \to -\infty} f(x) = f_{-\infty}$ and $\lim_{x \to \infty} f(x) = f_{\infty}$ exist. The Θ-function can then be obtained as

$$\Theta(x) = \lim_{\beta \to \infty} \frac{f(\beta x) - f_{-\infty}}{f_{\infty} - f_{-\infty}}. \qquad (5.75)$$

There exist, of course, infinitely many functions $f(x)$ with the desired property. Some familiar ones are $\arctan(x)$ or $\tanh(x)$; both functions are analytic around $x = 0$. However, their series expansions have a finite radius of convergence, which appears somewhat worrying given that we need to perform the limit $\beta \to \infty$. There are other choices, for instance, the probability integral $\Phi(x)$ or the sine integral $\mathrm{si}(x)$. Both functions are analytic on the entire real axis.

Let us assume that we have found a convenient function $f(x)$ with a series representation

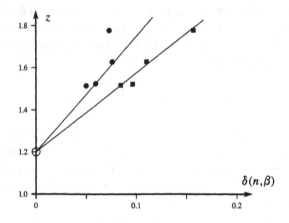

Figure 5.7. The dynamical critical exponent extrapolated to $\delta = 0$.

$$f(x) = \sum_{n=0}^{\infty} a_n x^n. \tag{5.76}$$

The Θ-function will then be given by

$$\Theta(x) = \lim_{\beta \to \infty} \lim_{n \to \infty} S(n, \beta x), \tag{5.77}$$

where we have introduced the function

$$S(n, \beta x) = \sum_{k=1}^{n} \theta_k (\beta x)^k. \tag{5.78}$$

The coefficients θ_k are readily calculated in terms of the coefficients a_k by use of (5.75).

We need a measure of how well the function $S(n, \beta x)$ approximates the Θ-function for a given choice of n and β (Ghaffari and Jensen 1996). Consider the integral

$$\delta(n, \beta) = \int_{-1}^{1} dx \, (\Theta(x) - S(n, \beta x))^2. \tag{5.79}$$

Any quantity calculated for given values of n and β can be considered a function of δ. In order to obtain the results pertinent to (5.72), we need to extrapolate to the limit $\delta \to 0$. Figure 5.7 shows how this procedure works in the case of the dynamical critical exponent. The values of z calculated for finite n depend on the representation $f(x)$ used in (5.75). However, both sets of data points for $z(\delta)$ can consistently be extrapolated to the value 1.2 at $\delta = 0$. The value $z \simeq 1.2$ is obtained from numerical simulations (Manna 1990, 1991a) as well as from the analytic approach known as the DDRG (see Section 5.5).

When looking at Figure 5.7 it is tempting to make a very enthusiastic jump in the seat. My advice is to jump and shout now, because it must sadly be pointed out that the quantitative success of Diaz-Guilera's calculation of the dynamical exponent, as judged from Figure 5.7, might be a bit accidental. The method involves an ε expansion around four dimensions, but there is no rigorous reason for the technique to work that well in two dimensions. Still, the achievements of this method are very encouraging. One of Diaz-Guilera's contributions was to show that the Zhang model defined in (5.70) is, in fact, in the same universality class as the BTW model defined in Section 4.2.2 (Diaz-Guilera 1994). See also (5.80) and (5.81) in the next section for $\alpha = 1/q_c$. This was expected from numerical simulations, but Diaz-Guilera added some rigor to this expectation.

5.4.2 Nonconservative Models

It will be interesting to apply Diaz-Guilera's procedure to the question concerning the criticality of nonconservative models of the type discussed in Sections 4.3 and 5.2.2 (Ghaffari et al. 1997). Let us introduce an element of nonconservation into the update of the Zhang model defined in (5.70). We do this in a way similar to the nonconservative update of the OFC model discussed in Section 4.3. That is, we replace the update in (5.70) with

$$E(\mathbf{r}, t+1) \to 0,$$
$$E(\mathbf{r}_{nn}, t+1) \to E(\mathbf{r}_{nn}, t) + \alpha E(\mathbf{r}, t). \tag{5.80}$$

When $\alpha < 1/q_c$ this update is nonconservative. Notice that, when this model is driven homogeneously, it is identical to the OFC model. In the update of the BTW model a fixed amount E_c, independent of the value of the field at the overcritical site, is distributed to the neighbor sites. This feature is incorporated in the following nonconservative BTW update:

$$E(\mathbf{r}, t+1) \to E(\mathbf{r}, t) - E_c,$$
$$E(\mathbf{r}_{nn}, t+1) \to E(\mathbf{r}_{nn}, t) + \alpha E_c. \tag{5.81}$$

It is easy to derive the continuum limit of the nonconservative BTW and the nonconservative Zhang model by writing the update in a form similar to (5.71). For the nonconservative BTW model we find

$$\partial_t E(\mathbf{r}, t) = [\alpha \tilde{D} \nabla^2 + (\alpha q_c - 1)]$$
$$\times \Theta(E(\mathbf{r}, t) - E_c)E_c + \eta_d(\mathbf{r}, t). \tag{5.82}$$

The nonconservation Zhang model is represented by the following equation:

$$\partial_t E(\mathbf{r}, t) = [\alpha \tilde{D} \nabla^2 + (\alpha q_c - 1)]$$
$$\times \Theta(E(\mathbf{r}, t) - E_c) E(\mathbf{r}, t) + \eta_d(\mathbf{r}, t). \quad (5.83)$$

Equations (5.82) and (5.83) were studied in the limit $\alpha = 1/q_c$ by Diaz-Guilera (1994). A mean field discussion of the models defined in (5.80) and (5.81) is given in Section 5.2.2.

We have introduced a source term $\eta_d(\mathbf{r}, t)$ to represent the driving of the system. Let us write this driving term in the form

$$\eta_d(\mathbf{r}, t) = \bar{\eta} + \eta(\mathbf{r}, t). \quad (5.84)$$

For models that are homogeneously driven, such as the OFC model (Section 4.3), the constant term $\bar{\eta}$ can be thought of as a representation of the homogeneous external drive between the avalanche updates.† The fluctuating part $\eta(\mathbf{r}, t)$ represents the fluctuations induced by the random initial condition. These spatial fluctuations in $E(\mathbf{r}, t = 0)$ will make different sites become over-critical at different times. For the randomly driven models, $\bar{\eta}$ is simply the average input per time and $\eta(\mathbf{r}, t)$ represents the fluctuations induced by the random addition of sand. Because the updating during the evolution of an avalanche is completely deterministic, Diaz-Guilera (1992) suggested that the appropriate noise in the Langevin equations is described by a correlator that is constant in time and delta-correlated in space:

$$\langle \eta(\mathbf{r}, t) \eta(\mathbf{r}', t') \rangle = 2\Gamma \delta^d(\mathbf{r} - \mathbf{r}'); \quad (5.85)$$

we assume $\langle \eta(\mathbf{r}, t) \rangle = 0$.

For now we shall concentrate on the BTW-like equation (5.82), although the same considerations can be made for the nonconservative Zhang model of (5.83). Neglect for a moment the fluctuating part of the drive; that is, let $\eta_d(\mathbf{r}, t) = \bar{\eta}$. Consider the limit of slow driving, $\bar{\eta} \leq (1 - \alpha q_c) E_c$. A moment's thought will convince the reader that any spatially homogeneous time-dependent solution $E(\mathbf{r}, t) = E(t)$ to (5.82) will evolve toward the time-independent solution $E(t) = \lim_{\varepsilon \to 0+} (E_c + \varepsilon)$. This observation leads us to look for solutions of the form

$$E(\mathbf{r}, t) = E_c + \delta E(\mathbf{r}, t), \quad (5.86)$$

where $\delta E(\mathbf{r}, t)$ has zero average.

To regularize the nonanalytical behavior of the step function, we consider (as mentioned previously) a real function $f : \mathbb{R} \to \mathbb{R}$ with the following properties:

---

† It must be added that it is not clear to what extent the homogeneously driven models can be represented by a Langevin equation with a noisy source term. The only source of disorder in (say) the OFC model is the random initial condition, and it may very well be that this disorder is badly represented by an additive noise term in the Langevin equation.

(a) $\lim_{x \to -\infty} f(x) = 0$;

(b) $\lim_{x \to \infty} f(x) = 1$; and

(c) the function can be expanded around $x = 0$ and has infinite radius of convergence.

The step function is represented by

$$\Theta(x) = \lim_{\beta \to \infty} f(\beta x). \qquad (5.87)$$

From (5.82), (5.86), and (5.87) we derive

$$\partial_t \delta E(\mathbf{r}, t) = \lim_{\beta \to \infty} [\alpha \tilde{D} \nabla^2 + (\alpha q_c - 1)] f(\beta \delta E(\mathbf{r}, t)) E_c + \eta_d(\mathbf{r}, t). \qquad (5.88)$$

Now imagine expanding $f(\beta \delta E)$ in a power series. The part of the equation independent of $\delta E(\mathbf{r}, t)$ can be removed from the equation by the following specific tuning of the constant part of the drive:†

$$\bar{\eta} = (1 - \alpha q_c) E_c. \qquad (5.89)$$

This may well represent the implicit tuning that takes place in simulations of the model. Recall that the model is not driven during the evolution of the avalanches. This allows the model to act at a point where the energy added to the model *between* the avalanches is precisely balanced by the energy dissipated *during* the avalanche. In the simulations of the model, dissipation takes place in the bulk when $\alpha < 1/q_c$, while dissipation always occurs at the open boundary of the system. Equation (5.89) represents the same type of balance.

Thus, we apply the RG analysis to the following equation:

$$\partial_t E(\mathbf{r}, t) = D \nabla^2 E + \lambda_3 \nabla^2 E^3 + \lambda_5 \nabla^2 E^5 + \cdots$$
$$+ \gamma_1 E + \gamma_3 E^3 + \gamma_5 E^5 + \cdots$$
$$+ \eta(\mathbf{r}, t), \qquad (5.90)$$

where $\langle E \rangle = \langle \eta \rangle = 0$. (To simplify notation we have dropped the δ in front of $E(\mathbf{r}, t)$; see (5.88).)

The Θ-function can be regularized by use of the probability integral

$$f(x) = \frac{1}{\sqrt{\pi}} \int_{-\infty}^{x} dt \, e^{-t^2}$$
$$= \frac{1}{2} + \frac{1}{\sqrt{\pi}} \sum_{k=1}^{\infty} (-1)^{k+1} \frac{x^{2k-1}}{(2k-1)(k-1)!}, \qquad (5.91)$$

in which case the coefficients in (5.90) are given by

† Imagine performing the limit $\delta E(\bar{r}, t) \to$ constant followed by $\beta \to \infty$.

$$D = \frac{\alpha \tilde{D} E_c \beta}{\sqrt{\pi}}, \qquad \gamma_1 = (\alpha q_c - 1)\frac{E_c \beta}{\sqrt{\pi}},$$

$$\lambda_3 = -\frac{\alpha \tilde{D} E_c \beta^3}{3\sqrt{\pi}}, \qquad \gamma_3 = -(\alpha q_c - 1)\frac{E_c \beta^3}{3\sqrt{\pi}},$$

$$\lambda_5 = \alpha \tilde{D} E_c \beta^5 10\sqrt{\pi}, \qquad \gamma_5 = (\alpha q_c - 1)\frac{E_c \beta^5}{10\sqrt{\pi}}.$$

Include successively more and more terms in (5.90). That is, first neglect all nonlinear terms, then include terms of order E^3, next include also terms of order E^5, and so forth. Study how the fixed point structure evolves as more and more terms are included.

The dynamical renormalization group (DRG) procedure consists of the following steps. Express all fields in Fourier space. Eliminate the short-wavelength contributions to the Fourier-transformed fields. Finally, rescale the k-vectors in Fourier space in order to recover the original Brillouin zone (see e.g. Medina et al. 1989). After this renormalization transformation in Fourier space, one obtains an equation similar to the starting equation with effective or renormalized coefficients. From the renormalized coefficients one can deduce the equations controlling the flow under renormalization in the parameter space.

The fixed points of the flow equations represent values of the renormalized coefficients for which the system will exhibit scale-invariant behavior in the hydrodynamical limit (large-distance and long-time behavior). The stability of the fixed points is calculated from a linear stability analysis. The idea is to study the nature of these fixed points as more and more nonlinearities are included in the equation of motion, that is, (5.90) or its equivalent for the Zhang model. The RG applied in the conservative case takes the form of an ε-expansion around the critical dimension $d_c = 4$.[†] Inclusion of the nonconservative terms introduces yet another critical dimension into the problem, namely, dimension 6. The RG technique applied to nonconservative systems is to be thought of as an ε-expansion about four dimensions supplemented by a perturbative expansion in the level of nonconservation given by the factor $(\alpha q_c - 1)$.

Before we explain the result for the nonconservative models, let us describe what happens in the conservative case. For the conservative model $\alpha = 1/q_c$ and therefore all $\gamma_i = 0$. Hence the system is described by

$$\partial_t E = D\nabla^2 E + \lambda_2 \nabla^2 E^2 + \lambda_3 \nabla^2 E^3 + \eta, \tag{5.92}$$

neglecting higher powers of E for the moment. From the work of Diaz-Guilera and Corral (Diaz-Guilera 1992; Corral and Diaz-Guilera 1997), we know that

---

[†] The value of the critical dimension is a consequence of the assumed form of the noise correlator in (5.85). For $\eta(r, t)$ delta-correlated in time as well as in space, $d_c = 2$.

this equation has no attractive fixed point when $\lambda_3 = 0$. For $\lambda_3 \neq 0$, however, an attractive fixed point does exist.

Now let us consider the conclusion of the RG analysis of the nonconservative models. No self-organized fixed point is found for any of the models, including the BTW model (5.82) and the Zhang model (5.83). The equations have a multitude of fixed points, but they are all repulsive in one or more directions in coupling space. No fixed point seems to become more attractive as more non-linearities are included in the equation of motion. That is, the nonconservative model will not, by itself, enter into the critical state; the self-organization is lost. This conclusion is in agreement with simulations (Ghaffari et al. 1997) as well as with the analytical analysis by Grinstein et al. (1990).

As the conservative model is approached by tuning $\alpha \to 1/q_c$, the characteristic length scale ξ of the noncritical nonconservative model diverges as $\xi \sim (1/q_c - \alpha)^{-\nu}$. It follows from the RG analysis of (5.82) and (5.83) that $\nu = 1/2$ in all dimensions (Ghaffari et al. 1997). This result is in excellent agreement with numerical results (Manna et al. 1990; Ghaffari et al. 1997) and in approximate agreement with the prediction of the dynamically driven renormalization group calculation to be discussed in the next section.

The homogeneously driven OFC model is yet another story. One can try to extract information about the critical behavior observed numerically for this model by tuning the fluctuations in the noise term to zero – that is, the limit $\Gamma \to 0$ in (5.85). At the moment, the outcome of this analysis is not completely understood.

The procedure developed by Diaz-Guilera of applying the renormalization group technique to the regularized Θ-function has shown how one can include in a systematic way the features of the threshold dynamics in a RG study, but there is a practical limitation to the applicability of this method. The handling of more and more nonlinearities in the equation of motion rapidly becomes prohibitively complex. Very large numbers of Feynman diagrams must be analyzed, and the resulting intractability is one problem. Another is that it is not altogether obvious to what extent the RG procedure works for nonlinear stochastic diffusion equations. The Kardar–Parisi–Zhang equation is one example of an equation that stubbornly resists the RG method (Esipov and Newman 1997).

In the next section we describe an alternative, and more direct, real-space renormalization analysis.

5.5 Dynamically Driven Renormalization Group Calculations

In the dynamically driven renormalization group (DDRG) approach, one analyzes directly the scale invariance of the spatiotemporal structures produced by the operation of the updating algorithm under consideration. The method developed by Pietronero and co-workers has been presented in a series of very readable and elegant papers (Pietronero and Schneider 1991; Pietronero et al. 1994; Vespignani, Zapperi, and Pietronero 1995; Erzan et al. 1995; Loreto et al. 1995; Loreto, Vespignanai, and Zapperi 1996). We shall illustrate the power of the procedure by discussing its application both to sandpile models and to the critical forest fire model. The discussion will be limited to two dimensions, although there are no problems (except for computational complexity) in applying the method to higher dimensions.

5.5.1 Renormalization Transformation

Consider a "height" model on a square lattice. To each site i is ascribed a real variable E_i. As before, we will call E_i the *energy* of site i. In the height models (BTW 1987), the condition for a site to become active depends only on the value of the dynamical field at that given site. This is the type of model we have discussed in Sections 4.2, 4.3, 5.2.1, and 5.3. For completeness we mention a related type of model – the "slope" model – where the update condition depends on the difference between more than one site. Accordingly, slope models are more complicated to analyze than height models. Moreover, slope models seem to have critical exponents that differ from those of height models (Kadanoff et al. 1989; Manna 1991a).

Here we concentrate on height models. As in the BTW sandpile, the energy of an overcritical site is transferred to a number of its neighbor sites. The idea is to study this redistribution of energy at different levels of coarse graining. One level of coarse graining, say level $k + 1$, is connected to the previous level k by lumping together a set of sites at level k in one larger site or cell of level $k + 1$ (see Figure 5.8).

The dynamics of the cells at level $k + 1$ is obtained from the collective overall behavior of the k-level sites constituting the $k + 1$ cell. The DDRG method operates in real space rather than Fourier space. In this respect, the method resembles the usual real-space renormalization group transformations as used, for instance, to analyze equilibrium-critical phenomena. However, the DDRG transformation integrates space and time in the analysis of the effect of the coarse graining on the dynamics of the system. In order to produce a closed set of equations by this procedure, we supplement the result of the coarse-grained analysis with the condition that the dynamics is stationary.

At the basic level, where the dynamics of the individual sites of the lattice is considered, the BTW model always distributes the energy of an overcritical site

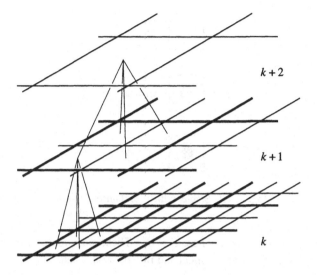

Figure 5.8. Coarse-grain analysis of the sandpile model via successive local averages.

to each of the four nearest neighbors. In the Manna model (Manna 1991b)†, the energy is always distributed to precisely two of the neighbor sites. In both cases, the number of neighbor sites that receive energy from an overcritical site will change when the analysis moves from one level of coarse graining to the next. Another quantity that will change during the coarse graining is the amount of energy added to a cell as a result of the external driving of the model. At level $k + 1$, the effective quantum of perturbation will be determined by the average amount of energy delivered by the random addition of energy packets to the subcells at level k.

Let us now describe the formalism needed to quantify these considerations (Pietronero et al. 1994; Vespignani et al. 1995). The sites of the lattice are divided into three different types according to their degree of stability with respect to the addition of δE, the quantum of perturbation (see Figure 5.9).

(1) *Stable* sites are those for which $E + \delta E < E_c$. Such sites, the open circles in Figure 5.9, remain stable even after receiving a quantum of the perturbation.

(2) *Critical* sites are those for which $E < E_c$ but $E + \delta E > E_c$. These sites, denoted by solid black circles, become unstable when perturbed.

† The two-state model constructed by Manna can be thought of as lattice gas of hard-core repulsive particles. Particles are placed at random on the lattice. If a lattice site becomes double occupied then both particles are removed from the site and placed in randomly chosen neighbor sites. The procedure is continued until single occupancy is reestablished; then a new particle is thrown onto the lattice.

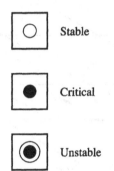

Figure 5.9. Stable, critical, and unstable sites.

(3) *Unstable* sites are those where $E > E_c$. These sites, represented by encircled black dots, are overcritical and ripe for relaxation.

The state of the system can be characterized by five variables $(\rho, \mathbf{P})$, where $\rho =$ the density of critical sites, $\mathbf{P} = (p_1, p_2, p_3, p_4)$, and

> $p_n =$ the probability that an unstable site distributes
> n packages of energy δE to precisely n nearest
> neighbor sites.   (5.93)

Obviously, $p_1 + p_2 + p_3 + p_4 = 1$. For the BTW model at the shortest length scale, $\mathbf{P} = (0, 0, 0, 1)$ since energy is always distributed to all four nearest neighbors. Similarly, for the model introduced by Manna, $\mathbf{P} = (0, 1, 0, 0)$. As the coarse graining proceeds, the vector $\mathbf{P}$ develops support on all four coordinates. One of the achievements of the DDRG is that it explicitly shows that the set $(\rho, \mathbf{P})$ in BTW and in the Manna model evolve to the same fixed-point values. In this way, DDRG verifies the expectation obtained from numerical simulations that the two models exhibit the same critical behavior.

At first sight it does not seem correct that the spatial properties of the model can be completely characterized by the spatially averaged density ρ alone. One would perhaps expect that we need to consider the spatial dependence of the field $\rho(\mathbf{r})$, and that the average $\rho = \langle \rho(\mathbf{r}) \rangle$ as well as correlation functions like $\langle \rho(\mathbf{r})\rho(\mathbf{r}') \rangle$ would have to be specified. This is partly true. We know from Dhar's work (see Majumdar and Dhar 1991 and Section 5.3) that the height–height correlation function $\langle E(\mathbf{r}) E(\mathbf{r}') \rangle - \langle E(\mathbf{r}) \rangle^2$ behaves like $|\mathbf{r} - \mathbf{r}'|^{-2d}$ for the hypercubic Abelian BTW model. These are algebraically decaying correlations and, as such, are normally said to be long-range correlations, which they certainly are if one compares them to exponentially decaying correlations. But if we compare the exponent of the decay, namely $-2d$, to the exponent of the decay $r^{-1/4}$ of the spin–spin correlation function at T_c for the two-dimensional Ising model (for a wonderfully lucid review see Kadanoff 1976), then the correlations in the two-dimensional BTW model are admittedly *much* weaker at long distances.

That the spatial correlations are in fact negligible in the critical state is indicated by a numerical result of Grassberger and Manna (1990). They considered the fluctuations σ_E in the total E-value of the lattice. The lattice consists of L^d sites. We have

$$\sigma_E = \left\{ \left\langle \left(\frac{1}{L^d} \sum_i E_i \right)^2 \right\rangle - \left\langle \frac{1}{L^d} \sum_i E_i \right\rangle^2 \right\}^{1/2}. \tag{5.94}$$

In their study of the BTW sandpile model, Grassberger and Manna found that σ_E scales with the size of the system as $\sigma_E \sim 1/L^{d/2}$. This is exactly what we obtain if there are no correlations between the values E_i on different sites. To see this, interpret the sum

$$S = \sum_{i=1}^{L^d} E_i \tag{5.95}$$

as the displacement of a random walker after $N = L^d$ statistically independent steps E_i. From the central limit theorem (Reif 1965, sec. 1.11) we know that $\langle S^2 \rangle - \langle S \rangle^2 \sim \sqrt{N}$. Divide this equation by $N^2 = L^{2d}$ and take the square root; this leads to the result $\sigma_E \sim 1/L^{d/2}$.

During the evolution of an avalanche, spatial correlations do build up. We saw from Dhar's Δ-matrix formalism (see Section 5.3) that the toppling correlations decay much slower (like r^{2-d}) than equal-time height–height correlations (which decay like r^{-2d}). The DDRG procedure takes the dynamical correlations into account through the renormalization transformations that describe the relaxation dynamics. The random action of different avalanches tends to wash out the correlations built up by the individual avalanches. That is, when averaged over many avalanches, $\langle \rho(\mathbf{r})\rho(\mathbf{r}') \rangle$ will exhibit only weak correlations. One can therefore assume that the "medium" through which avalanches evolve is well characterized by $\rho = \langle \rho(\mathbf{r}) \rangle$ only.

The DDRG transformation consists of an analysis of how $(\rho, \mathbf{P})$ and the perturbation δE are transformed as the scale of the coarse graining is changed. We now describe how the link between the different levels of coarse graining is established. Let us double the linear size of the cell in each step (see Figure 5.8). The process at level $k + 1$ corresponds to scale b, and level k corresponds to scale $b/2$. The definitions of *stable, critical,* and *unstable* must be generalized to the kth level of coarse graining. A cell at level $k + 1$ is defined as

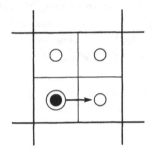

Figure 5.10. Relaxation inside a level-$(k + 1)$ cell. The cell is considered
stable at level $k + 1$.

stable if the addition of $\delta E(b)$, the quantum of perturbation pertinent to level
$k + 1$, does not lead to energy being transported into the neighboring cells of
linear size b. Relaxations can occur inside a stable cell as long as no energy is
transported across the boundary of the square $b \times b$ cell (see Figure 5.10). On
the other hand, the cell is *critical* if energy is transported into a number of the
neighboring cells. The relaxation probabilities p_i^{k+1} at level $k + 1$ can be expressed in terms of p_i^k. In order to derive the relation between these two sets of
relaxation probabilities we must study how a cell at level $k + 1$ relaxes. This
relaxation depends on the number $\alpha = 1, 2, 3, 4$ of critical k-cells inside the
$(k + 1)$-cell.

The case $\alpha = 1$ need not be considered, since we are concerned only with
spatially connected avalanches. We ensure this restriction by including in the
relaxation analysis only those cells for which the internal relaxation can span
across the linear extent of the cell (see Figure 5.11).

Cells at level $k + 1$ that contain $\alpha = 2$ critical k-cells can contribute to p_i^{k+1}
for all $i = 1, 2, 3, 4$. In order not to be completely overwhelmed by computational details, we will restrict our discussion to the calculation of p_1^{k+1} (see
Figure 5.12). The requirement that the considered process can connect across
the cell implies that a relaxation at level $k + 1$ always involves two successive
relaxations at level k. For example, a p_1 process can be followed by either a p_1,
a p_2, or a p_3 process (see Figure 5.13). A p_1 process that is followed by a p_4
process contributes not to the renormalization of p_1^{k+1} but rather to the renormalization of p_2^{k+1}. The p_1–p_1 process depicted in Figure 5.13 contributes to
p_1^{k+1} by a term $(\frac{1}{4}p_1)(\frac{2}{4}p_1)$. The first factor is the probability that the unstable site relaxes through a p_1 process in the direction of the critical site within
the cell. Since the processes are assumed to be isotropic, this specific direction is picked with probability $\frac{1}{4}$. The second factor is the probability that the

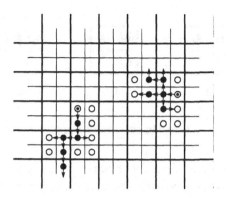

Figure 5.11. Processes involving $\alpha = 1$ cells.

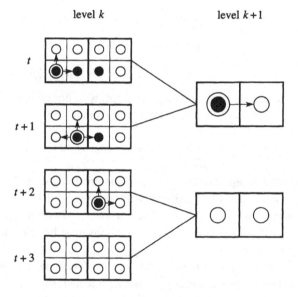

Figure 5.12. Chain of k-level processes contributing to level-$(k + 1)$ processes.

subsequent relaxation takes place through a p_1 process pointing in one of the two directions over the edge of the cell – hence the factor $\frac{2}{4}$. We consider the p_1–p_2 double process sketched in Figure 5.13 as a second example. The first process contributes a factor $\frac{1}{4}p_1$ as before. The second process is a p_2 process

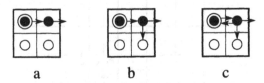

Figure 5.13. p_1 process followed by either another p_1 process or a p_2 or a p_3 process.

where precisely one relaxation takes place over the edge of the cell and other relaxations occur within the cell. There are $K_{4,2} = 6$ ways of choosing the two directions of relaxation, only four of which lead to processes that contribute to the renormalization of p_1^{k+1}; that is, the second process is characterized by a factor $\frac{2}{3}p_2$. The total contribution to p_1^{k+1} from $\alpha = 2$ configurations is given by

$$p_1^{k+1}(\alpha = 2) = (\tfrac{1}{4}p_1^k + \tfrac{1}{6}p_2^k)(\tfrac{1}{2}p_1^k + \tfrac{2}{3}p_2^k + \tfrac{1}{2}p_3^k)$$
$$+ (\tfrac{1}{6}p_2^k + \tfrac{1}{4}p_3^k)(\tfrac{1}{2}p_1^k + \tfrac{1}{6}p_2^k) \qquad (5.96)$$
$$+ (\tfrac{1}{6}p_2^k + \tfrac{1}{4}p_3^k)(\tfrac{3}{4}p_1^k + \tfrac{1}{2}p_2^k + \tfrac{1}{4}p_3^k).$$

In a similar (though clearly much more involved) way, one derives expressions for $p_i^{k+1}(\alpha = 2)$ for $i = 2$, 3, and 4. Finally, we impose the normalization

$$\sum_{i=1}^{4} p_i^{k+1}(\alpha = 2) = 1. \qquad (5.97)$$

The procedure is repeated for configurations involving $\alpha = 3$ and $\alpha = 4$ critical sites, and the renormalized probability at level $k + 1$ is obtained by averaging over the configurations of different α-values. To do this, the statistical weight $W_\alpha(\rho^k)$ of the various α-configurations must be calculated; this is discussed in Appendix E.

Putting all these elements together leads to the functional relationship between (a) the process probabilities p_i^{k+1} at length scale b and (b) the probabilities p_i^k and the density of critical sites ρ^k at length scale $b/2$:

$$p_i^{k+1} = \sum_{\alpha=2}^{4} W_\alpha(\rho^k)p_i^{k+1}(\alpha)$$
$$\equiv \mathcal{F}_i(p_1^k, \ldots, p_4^k, \rho^k). \qquad (5.98)$$

The explicit form of the functions $\mathcal{F}_i$ is rather tedious to write out. (Vespignani et al. 1995); (5.96) and (5.98) give an idea of the calculations involved.

The scale-invariant behavior of the model is described by the fixed-point solution $p_i^k = p_i^{k+1} = \mathcal{F}_i(p_1^k, \ldots, p_4^k, \rho^k)$ for $i = 1, \ldots, 4$. In order to solve the fixed-point equation we need to establish a coupling between the probabilities p_i^k and the density ρ^k. This amounts to linking the dynamical properties and transition probabilities to the spatial characteristics. The relation is easily obtained by noting that – in the stationary state (the situation of interest) – there is, on average, a balance between the amount of energy flowing into a region and the energy flowing out of that region. Note that we have conservative models in mind, not the dissipative models of the OFC type (see Sections 4.3 and 5.2.2). This balance is expressed by the following equation:

$$\delta E(b/2) = \rho^k [\delta E(b/2) p_1^k + 2\delta E(b/2) p_2^k$$
$$+ 3\delta E(b/2) p_3^k + 4\delta E(b/2) p_4^k]. \tag{5.99}$$

The left-hand side denotes the amount of energy added to a cell of length scale $b/2$ in a perturbation at coarse-grained level k. The right-hand side represents the probability ρ^k that the site becomes overcritical and therefore relaxes through one of the four possible channels of relaxation. The four terms within brackets represent the amount of energy transported out of the $b/2$ cell through the different modes of relaxation multiplied by the individual probabilities for the respective processes. The energy δE drops out of the equation, and we obtain an expression for ρ in terms of the probabilities:

$$\rho^k = [p_1^k + 2p_2^k + 3p_3^k + 4p_4^k]^{-1}. \tag{5.100}$$

The renormalization of the dynamical evolution of the avalanche is characterized by (5.98) and (5.100). The behavior at long length scales is obtained by repeated iteration of the transformation in (5.98). For any initial value of the pentuple $(\rho, \mathbf{P})$, the iteration flows toward one globally attractive fixed point $(\rho^*, \mathbf{P}^*)$ as $k \to \infty$. Values of the fixed-point parameters, as well as evolution observed under iteration in the Manna two-state and the BTW four-state model, are listed in Table 5.1.

Thus, any conservative sandpile-like model that can be defined in terms of the transition probabilities p_i is controlled by the same fixed point and therefore exhibits the same scale-invariant behavior. Because this class of models is described by a fixed point that is attractive throughout the parameter space, the critical scale-invariant behavior will always be present. No control parameter has to be tuned in order to put the model at the critical point, so these models are truly self-organized critical.

Table 5.1. *Evolution of the parameter
vector* $(\rho, \mathbf{P})$ *under iteration of
the renormalization transformation*

k	ρ	p_1	p_2	p_3	p_4
Manna two-state model					
0	0.1	0	1	0	0
1	0.612	0.436	0.495	0.068	0.001
2	0.575	0.405	0.463	0.118	0.013
3	0.542	0.362	0.456	0.158	0.024
4	0.518	0.324	0.434	0.188	0.033
∞	0.468	0.240	0.442	0.261	0.057
BTW four-state model					
0	0.9	0	0	0	1
1	0.252	0	0	0.033	0.967
2	0.308	0	0.012	0.726	0.262
3	0.353	0.030	0.261	0.553	0.152
4	0.388	0.090	0.357	0.437	0.116
∞	0.468	0.240	0.442	0.261	0.057

Source: Data from Vespignani et al. 1995.

5.5.2 Exponents

The exponent of the avalanche size distribution $P(s) \sim s^{-\tau}$ can be calculated when the critical point $(\rho^*, \mathbf{P}^*)$ is known. The DDRG transformation does not prove that the avalanche sizes are distributed according to a power law, although it does show explicitly the existence of a fixed point of the scale transformation (see (5.98)). This fixed point would *not* exist if the model possessed a characteristic length scale. To see this, assume the opposite – namely, that the model is characterized by a specific length scale, say a specific size l_0 of the avalanches. The change of scale studied in the previous section would then always be relevant. The behavior of the coarse-grained system will depend on the ratio b/l_0. One type of behavior will exist for $b < l_0$ and another for $b > l_0$. This will make p_i^k change in each step, and no fixed point would exist. However, if $l_0 = \infty$ then the only length scale of the model is the microscopic lattice spacing a_0, and as the scale b becomes large compared to a_0 we expect the dependence on a_0 to diminish gradually and the fixed-point behavior to emerge.† Having verified the scale invariance of the

---

† In principle, the fixed point could also be consistent with $l_0 = 0$. This corresponds to decoupled, completely uncorrelated behavior of the lattice sites. From the definition and the simulations of the model we certainly do not expect this type of behavior.

model, one can safely assume that event distributions follow power laws, the functional form that is consistent with the lack of a characteristic scale (see Section 2.1).

We assume that $P(s) \sim s^{-\tau}$ and then determine τ from the dynamical properties at the critical point $(\rho^*, \mathbf{P}^*)$. The size s of an avalanche is connected to its linear extension r through its fractal dimension D: $s \sim r^D$. In two dimensions, the avalanches of the sandpile models are compact; that is, $D = 2$. This is clearly seen in numerical simulations (Christensen et al. 1991) and was shown by use of the DDRG by Pietronero and Schnieder (1991). We can therefore write

$$P(r)\,dr \sim P(s)\,ds \sim r^{(1-2\tau)}\,dr, \tag{5.101}$$

since $ds = 2r\,dr$. Here $P(r)\,dr$ denotes the probability that an avalanche has linear size between r and $r + dr$. Let us label as r_1-*avalanches* those of linear extent larger than r_1. Consider the probability K_{r_1,r_2} that an r_1-avalanche does not reach beyond r_2. We have

$$K_{r_1,r_2} = \frac{\int_{r_1}^{r_2} P(r)\,dr}{\int_{r_1}^{\infty} P(r)\,dr} = 1 - \left(\frac{r_2}{r_1}\right)^{2(1-\tau)}. \tag{5.102}$$

Next, consider again the scale transformation discussed in Section 5.5.1. Imagine a coarse graining in terms of cells of size b. We can readily calculate the probability $K_{b,2b}$ that an avalanche at this level of coarse graining involves precisely one update. The single update can consist of a process involving relaxation into 1–4 of the neighbor cells. Each neighbor cell must be a stable cell; otherwise, the relaxation process will continue beyond a single update. The total probability that a single update avalanche occurs is accordingly

$$K_{b,2b} = p_1(1 - \rho) + p_2(1 - \rho)^2 + p_3(1 - \rho)^3 + p_4(1 - \rho)^4, \tag{5.103}$$

where p_i and ρ correspond to scale b. However, in the critical state, p_i and ρ become scale-invariant (at large scales) and we evaluate $K_{b,2b} = 0.2966$ by substituting the fixed-point values $(\rho^*, \mathbf{P}^*)$ from Table 5.1 into (5.103). The avalanches involved in a one-step process at level b correspond to avalanches of linear extent between $r_1 = b$ and $r_2 = 2b$. From (5.102) we can then calculate the exponent τ as follows:

$$K_{b,2b} = 1 - 2^{2(1-\tau)}$$

$$\Downarrow \tag{5.104}$$

$$\tau = 1 - \frac{\ln(1 - K)}{2\ln 2} = 1.253.$$

This value is in remarkably good agreement with the numerical results obtained by Manna: $\tau = 1.28$ for the two-state model (Manna 1991b) and $\tau = 1.22$ for the BTW model (Manna 1990, 1991b). It is also very close to the result $\tau = 5/4$ obtained by Priezzhev et al. (1996) by an extension of Dhar's Δ-matrix formalism (see Section 5.3).

The dynamical critical exponent z can also be calculated from the DDRG formalism. The exponent z relates spatial scale r to time scale t through $t \sim r^z$. Recall, for instance, a random walker. The average square distance r^2 moved after t time steps scales as $r^2 \sim t$; that is, the random walk is characterized by a dynamical exponent $z = 2$. Likewise, we can ask about the relation between the number of time steps or updates t involved in an avalanche of spatial linear extent r. For a scale-invariant system we again expect the relation between t and r to be a simple power $t \sim r^z$. The exponent z can be calculated via DDRG if one can calculate the average number of updates or time steps t_b involved in a relaxation process at scale b. By definition, $t_b \sim b^z$. The renormalization transformation can be used to find the relation between t_b and t_{2b}. The time scale at coarse-grained level $2b$ will be equal to t_b times the average number of processes N_b at level b involved in a level-$2b$ relaxation. That is,

$$t_{2b} = N_b t_b \Rightarrow z = \frac{\ln(t_{2b}/t_b)}{\ln 2} = \frac{\ln N_b}{\ln 2}. \tag{5.105}$$

The factor N_b is calculated by the same procedure as was used in Section 5.5.1 to calculate the transformation of the transition probabilities p_i; see (5.96). Each of the subprocesses contributing to the renormalization of $p_1(2b)$ starting from $\alpha = 2$ configurations involved exactly two sub-updates at scale b – for example, a $p_1(b)$ process followed by another $p_1(b)$ process or a $p_1(b)$ process followed by a $p_2(b)$ process, and so on. All the $p_1(2b)$ processes not starting from $\alpha = 2$ configurations (or $p_i(2b)$, $i = 2, 3, 4$ processes) can involve, at the b-level, more than two individual $p_i(b)$ processes. The factor N_b is obtained by calculating the average number of individual (nonsimultaneous) $p_i(b)$ processes needed. One finds that $N_b = 2.247$ and therefore, from (5.105), that $z = 1.168$. The dynamical critical exponent found by Manna in his large-scale numerical simulations is $z \simeq 1.2$.

The scaling relations discussed in Section 4.2 can be used to determine other exponents characterizing the event distributions of the model in terms of τ and z.

Exercise. Note that $z = \gamma_3 = \gamma_2/\gamma_1$ in the notation of Section 4.2.4. Make use of the compactness $\gamma_2 = 2$ of the avalanches in two dimensions and the scaling relations of Section 4.2.4 to derive the relations

$$\lambda = 2\tau - 1, \tag{5.106}$$

$$\alpha = 1 + 2\frac{\tau - 1}{z}. \tag{5.107}$$

5.5.3 Nonconservative Models

One of the crucial questions concerning SOC is how robust the critical and/or self-organizing behavior is with respect to alterations of the microscopic update defining the model. One of the more interesting possible alterations to the dynamics of a model is to make it nonconservative. We have previously considered the effect of lack of conservation (in particular, in Sections 4.3 and 5.2.2). Here we want to describe how a degree of dissipation can be incorporated in the DDRG formalism. When a site becomes overcritical and distributes energy to its neighbor sites, energy can be forced to leave the system in many different ways (Manna 1990; Christensen 1992; Olami et al. 1992; Lise and Jensen 1996; Ghaffari et al. 1997).

An elegant possibility is to introduce completely dissipative updates every so often but otherwise leave the algorithm unchanged (Vespignani et al. 1995). An overcritical site is updated according to the probabilities p_n that the amount $n\delta E$ of energy is removed from the site; see (5.93). With probability γ this energy is simply thrown out of the system (i.e., dissipated); with probability $(1 - \gamma)$ the energy is distributed to n of the neighbor sites, each receiving a packet of energy equal to δE. Hence, every $1/\gamma$ update leads to a loss of energy, and all other updates are strictly conservative.

The expressions for the renormalization of the p_n probabilities, equations like (5.96) and (5.98), are modified. The dissipation parameter γ will enter on the right-hand side of, say, (5.96). Consider for example the first term on the right-hand side of (5.96). This term, $(p_1/4)(p_1/2)$, describes a p_1 process at level $b/2$ followed by yet another p_1 process. Dissipation will modify this term as follows:

$$\frac{p_1}{4}\frac{p_1}{2} \to (1 - \gamma)\frac{p_1}{4}(1 - \gamma)\frac{p_1}{2}. \tag{5.108}$$

The two factors $(1 - \gamma)$ are the probabilities that the p_1 processes actually transfer energy to neighbor sites and thereby contribute to the considered p_1 process at level b. If, say, the second p_1 process at level $b/2$ were purely dissipative (which happens with probability γ), then no energy would be transferred out of the b-cell and the process would not contribute to the p_1 process at level b. In general, subprocesses at level $b/2$ that contribute to the renormalization of the p_n processes at level b can be a mixture of, say, m purely dissipative (probability γ^m) and, say, q completely conservative (probability $(1 - \gamma)^q$) subprocesses – as long as the chain of subprocess leads to a transfer of energy across the edge of the b-cell. An additional factor $\gamma^m (1 - \gamma)^q$ must be multiplied into the weight of the process in the renormalization equations similar to (5.96).

The dissipation level γ is also renormalized as one changes the length scale. At coarse-graining level b, the dissipation level $\gamma(b)$ is defined as the ratio

$$\gamma(b) = \frac{\Delta E_{\text{lost}}(b)}{\Delta E_{\text{lost}}(b) + \Delta E_{\text{tran}}(b)} \tag{5.109}$$

between the average amount of energy lost ΔE_{lost} due to dissipation and the average total energy (dissipated plus transferred energy ΔE_{tran}) involved in the update of a b-level cell. The quantities ΔE_{lost} and ΔE_{tran} can be calculated by averaging the energy loss and transfer over the individual chains of $b/2$ processes constituting the update of the b-level processes.

The coupling between ρ and p_i is unaffected by dissipation. This is because the equation for the energy balance, (5.99), is replaced by the sums of energy-transferring processes and energy-dissipating processes:

$$\delta E(b) = (1 - \gamma(b))\rho(b) \sum_{n=1}^{4} n\delta E(b) p_n(b)$$

$$+ \gamma(b)\rho(b) \sum_{n=1}^{4} n\delta E(b) p_n(b)$$

$$= \rho(b) \sum_{n=1}^{4} n\delta E(b) p_n(b). \tag{5.110}$$

Hence ρ is once again eliminated from the right-hand side of the RG equations by use of (5.100).

The renormalization procedure leads to a set of equations of the form

$$\gamma^{k+1} = G(\gamma^k, p_1^k, \ldots, p_4^k),$$
$$p_i^{k+1} = \mathcal{F}_i(\gamma^k, p_1^k, \ldots, p_4^k). \tag{5.111}$$

This set of equations has two fixed points $(\gamma^*, \mathbf{P}^*)$. One fixed point is at $\gamma^* = 0$ and the other is at $\gamma^* = 1$. Both fixed points have the same set of coordinates along the p_i directions in parameter space, namely the fixed-point values of p_i^* found in Section 5.5.1 (see Table 5.1). The conservative fixed point at $\gamma^* = 0$ is repulsive, whereas the completely dissipative fixed point at $\gamma^* = 1$ is attractive (see Figure 5.14). This behavior follows from the stability analysis of the linearized RG equations. The lack of conservation for $\gamma \neq 0$ introduces a length scale ξ. At the fixed point $\gamma^* = 0$ this length is infinite. Let us outline

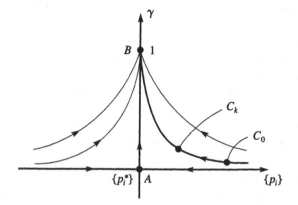

Figure 5.14. The renormalization flow in the vicinity of a repulsive fixed point.

how the RG flow close to the repulsive fixed point enables us to determine the exponent ν of the characteristic length scale

$$\xi \sim \gamma^{-\nu}. \qquad (5.112)$$

The procedure is standard for repulsive fixed points like those encountered in equilibrium critical phenomena (Binney et al. 1992). We repeat the arguments here in our special case for completeness.

The RG equations in (5.111) are linearized about the repulsive fixed point at $\gamma = 0$. The linearized iterative equations have the form

$$\begin{pmatrix} \gamma \\ \delta p_1 \\ \delta p_2 \\ \delta p_3 \\ \delta p_4 \end{pmatrix}_{k+1} = \begin{pmatrix} \lambda & 0 & 0 & 0 & 0 \\ \cdot & & & & \\ \cdot & & M & & \\ \cdot & & & & \\ \cdot & & & & \end{pmatrix} \begin{pmatrix} \gamma \\ \delta p_1 \\ \delta p_2 \\ \delta p_3 \\ \delta p_4 \end{pmatrix}_{k} , \qquad (5.113)$$

where $\delta p_i = p_i - p_i^*$. The matrix on the right-hand side is the Jacobian matrix of the transformation given in (5.111) evaluated at the point $(\gamma, \mathbf{P}) = (0, \mathbf{P}^*)$. Hence,

$$\lambda = \left[\frac{\partial G}{\partial \gamma} \right]_{\gamma=0}. \qquad (5.114)$$

It is easy to see why all four elements in the top row of the matrix in (5.113) are zero. The scale-dependent dissipation parameter $\gamma(b)$ defined in (5.109) is of order $\gamma(b/2)$. This is because all terms contributing to ΔE_{lost} are at least of order $\gamma(b/2)$, since at least one of the subprocesses entering the chain of processes contributing to the loss of energy must be a dissipation process. The top-row elements are given by $\partial G/\partial p_i$ for $i = 1, \ldots, 4$ evaluated at $\gamma = 0$. The function G is defined by the right-hand side of (5.109). Hence $\partial G/\partial p_i \sim \gamma$, and these elements vanish for $\gamma = 0$. The matrix M in (5.113) is the Jacobian of the function $\mathcal{F}$ evaluated at $\gamma = 0$ and p_i set equal to the fixed-point values already considered for the conservative case in Section 5.2.1. Accordingly, we know that this matrix is the stability matrix of an attractive fixed point in the parameter space $(p_1, \ldots, p_4)$. All eigenvalues of the matrix are therefore smaller than unity. The five eigenvalues of the matrix in (5.113) are obviously given by the element λ and the four eigenvalues of the matrix M. The detailed calculation gives $\lambda = 2.81$.

Let us now determine the exponent ν. Consider the flow diagram in Figure 5.14. The correlation length is infinite precisely at the fixed point A, which is why the change of scale leaves $(\gamma, \mathbf{P})$ unchanged for this choice of parameters. At the fixed point B, the parameter set $(\gamma, \mathbf{P})$ is also unaffected by the scale transformation. The insensitivity to length scale at this fixed point is, however, entirely different from the lack of scale dependence encountered at point A. At point B, the dissipation parameter is $\gamma = 1$. There is no interaction between neighboring sites, and each cell relaxes independently of the surroundings. For this reason it makes no difference if we consider cells of size $b/2$ or cells of size b, since there is no communication between the cells in either case. Here the correlation length $= 0$. For any other location of the parameter set $(\gamma, \mathbf{P})$ in the phase space depicted in Figure 5.14, the effective set of parameters $(\gamma(b), \mathbf{P}(b))$ will change during coarse graining from b to $2b$ and so forth. This change in $(\gamma(b), \mathbf{P}(b))$ reflects the relative change between the cell size b and the coherence length $\xi(\gamma)$.

Imagine that we start at a cell size equal to a_0 (the lattice spacing of our model) characterized by the parameter set $C_0 = (\gamma(a_0), \mathbf{P}(a_0))$ located in the phase space at point C_0 in Figure 5.14. We now perform successively the change of cell size

$$b_0 \to 2a_0 \to \cdots \to b_k = 2^k a_0 \to \cdots .$$

As we consider larger and larger scales, the parameter set $C_k = (\gamma(b_k), \mathbf{P}(b_k))$ flows along the trajectory passing through the point C_0. The points C_k initially flow toward the fixed point at A but then, after a certain number of iterations, "take off" toward the attractive trivial fixed point located at B. The cell size b_k will at first be smaller than the correlation length $\xi(\gamma(a_0))$ of the model for some given small value $\gamma(a_0) \equiv \gamma_0$ of the dissipation parameter. As long as $b_k < \xi(\gamma_0)$, the model behaves under scale transformation essentially as if $\xi = \infty$ and the flow of the points in C_k will be more or less in the direction of A.

The effective dissipation parameter $\gamma(b_k)$ will change very little from iteration to iteration. However, after a number (say, k_t) of iterations, the cell size has increased to become equal to the correlation length:

$$b_{k_t} = \xi_0 \Rightarrow \xi(\gamma_0) = 2^{k_t} a_0. \tag{5.115}$$

As the iteration is continued, the relation between the cell size and the correlation length becomes more like the situation at the fixed point B than at fixed point A, and the flow of C_k will take off toward point B. The effective dissipation parameter $\gamma(b_{k_t})$ characterizing the model at the length scale b_{k_t}, where the flow takes off toward the trivial fixed point, is related to the dissipation parameter at scale a_0 as

$$\gamma(b_{k_t}) = \lambda^{k_t} \gamma(a_0); \tag{5.116}$$

this follows from (5.113). We can use (5.116) to express k_t in terms of γ_0, $\gamma_t \equiv \gamma(b_{k_t})$, and the eigenvalue λ:

$$k_t = \ln(\gamma_t/\gamma_0)/\ln(\lambda).$$

From (5.115) it now follows that

$$\xi(\gamma_0) = a_0 \left[\frac{\gamma_t}{\gamma_0}\right]^{(\ln 2)/(\ln \lambda)}. \tag{5.117}$$

We conclude that $\xi \sim \gamma_0^{-\nu}$ where $\nu = \ln 2/\ln \lambda$.

Substituting the numerical value $\lambda = 2.81$ leads to $\nu = 0.67$. In simulations of a similar nonconservative model, $\nu \simeq 0.5$ (Manna et al. 1990; Ghaffari et al. 1997), in agreement with renormalization group analysis of Langevin equations (see Section 5.4.2). See also Section 4.3 for more details concerning simulations of various types of nonconservative models.

The result that the fixed point at $\gamma = 0$ is repulsive is in agreement with our previous conclusion, derived from simulation studies (see Section 4.3) and RG analysis of Langevin equations (see Section 5.4), that critical behavior of a model survives the introduction of a degree of nonconservation only if the model is driven homogeneously and the updating algorithm couples the overcritical site to the neighbor sites in an essential way – namely, by letting the amount of energy distributed depend on the amount of energy present at the overcritical site. The model analyzed here via DDRG distributes a fixed amount of energy (a multiple of δE) that is independent of the value of $E > E_c$ at the overcritical site. Our mean field argument in Section 4.3 then suggests that the branching ratio < 1, and that the model is therefore noncritical, as soon as dissipation is introduced.

5.5.4 Forest Fire Models

The DDRG transformation has also been applied to the forest fire model described in Sections 4.5 and 5.2.4 (see Loreto et al. 1995). The technique is very similar to the procedure outlined in Sections 5.5.1 and 5.5.2. The scale transformation is applied to the five-dimensional parameter space consisting of the growth probability p, the ratio $\Theta = f/p$ between the ignition probability f and the growth probability, and the three densities $\vec{\rho} = (\rho_e, \rho_t, \rho_f)$, where ρ_e is the density of empty sites, ρ_t the density of trees, and ρ_f the density of burning trees. The DDRG technique demonstrates that a fixed point exists at $p^* = 0$, $\Theta^* = 0$, and $\vec{\rho} = (\frac{2}{3}, \frac{1}{3}, 0)$. That is, critical behavior will occur in the limit of slowly growing trees, $p \to 0$. Though slow-growing, the trees must grow quickly compared to the time interval between spontaneously ignited fires, $\Theta = f/p \to 0$. The fixed point is repulsive along the direction of the parameter Θ. Hence, the model is not self-organizing through its own intrinsic dynamics to the critical point. Criticality is observed only when the parameter Θ is moved toward the limit $\Theta = 0$ by external tuning.

As before, the exponent controlling the divergence of the characteristic length in the model can be calculated from the eigenvalue of the stability matrix of the RG equation. The result is that the average radius of clusters of growing trees diverges as $R \sim \Theta^{-\nu}$ while the critical point is approached, $\Theta \to 0$. The numerical value of the exponent ν depends on the size of the subcell used in the coarse-graining procedure. For cells consisting of 2×2 subcells, $\nu = 0.73$; an increase to 3×3 cells changes the estimated value of the exponent to $\nu = 0.65$. The numerical estimate is $\nu \simeq 0.58$ (Grassberger 1993; Clar et al. 1994).

At the critical point, the distribution of forest clusters is a power law $P(s) \sim s^{-\tau}$. The exponent τ can be derived from the DDRG calculation by arguments similar to those presented in Section 5.5.2. The result is $\tau = 1.16$, in nice agreement with the simulations (Grassberger 1993).

Note Added in Proof. Two recent preprints (M.-L. Chabanol and V. Harkim; H.-M. Bröker and P. Grassberger) discuss the random neighbor OFC model in great analytic detail. Both papers conclude that the model is noncritical for $\alpha < 1/q_c$. The average avalanche size is found to diverge *exponentially* when $\alpha \to 1/q_c$. This suggests that the conclusion of Section 4.3.3 and Section 5.2.2 (see equation (5.15)) is incorrect.

6

Is It SOC or Not?

Ever since the term "self-organized criticality" was introduced by Bak, Tang, and Wiesenfeld (BTW) in their 1987 paper for *Physical Review Letters,* the concept has been surrounded by a hectic air of controversy. There are a number of reasons for this. One reason is the bold and optimistic claims that were made. The attitude was that here finally is a line of thinking that will allow us to bring the statistical physics of Boltzmann and Gibbs in touch with the exciting real world of nonequilibrium physics, and that SOC is powerful enough to explain everything from mountain formation to stock-market variation. Supergeneral theories always meet a certain amount of skepticism from expert scientists working in the specific fields. It is difficult to draw a precise line between the general and the specific. It might not appear likely to the geologist that the many specific details of earthquakes can be understood in terms of a simple numerical cellular automaton. The biologist working on the immensely complicated interconnected web of evolving species might not find it anything but a bad joke to represent evolution in terms of a string of random numbers with nearest neighbor interaction only.

So what, then, is SOC good for? Let us consider some important questions.

(1) Can we identify SOC as a well-defined distinct phenomenon different from any other category of behavior?
(2) Can we identify a certain construction that can be called a *theory* of self-organized critical systems?
(3) Has SOC taught us anything about the world that we did not know prior to BTW's seminal 1987 paper?
(4) Is there any predictive power in SOC – that is, can we state the necessary and sufficient conditions a system must fulfill in order to exhibit SOC? And, if we are able to establish that a system belongs to the category of SOC systems, does that then actually help us to understand the behavior of the system?

With some caution I think it is meaningful to answer all these questions in the affirmative. In this chapter we shall discuss in what sense SOC has been a success.

Originally, self-organized criticality was suggested to be *the* typical behavior of interacting many-body systems. The abundance of fractal structures, temporal as well as spatial, was imagined to be an effect of a generic tendency – pertinent to most many-body systems – to develop by themselves into a critical scale-invariant state.

At one level the answer is clear. Certainly not all systems that organize themselves into one specific state will, when gently driven, exhibit scale invariance in that self-organized state. The experiments on sandpiles are a prime example (see Section 3.2). Neither is all observed power law behavior an effect of dynamical self-organization into a critical stationary state. The work by Sethna and co-workers on the Barkhausen noise (Sethna et al. 1993; Perković, Dahmen, and Sethna 1995) is a very appealing specific example of what Sornette (1994) has called "power laws by sweeping of an instability." Despite these caveats, it certainly seems possible to formulate a working definition of SOC systems.

6.1 Where Is SOC to Be Found?

We will expect SOC behavior in *slowly driven, interaction-dominated threshold* systems. It may thus be useful to introduce a new abbreviation – SDIDT systems – in order to emphasize these essential dynamical ingredients. We could call this a *constructive* definition, in contrast to the usual term, SOC, which stresses the behavior of the system. If a system exhibits power laws without any apparent tuning then it is said to exhibit self-organized criticality; SOC is a *phenomenological* definition rather than a constructive one.

The notion of an "interaction-dominated threshold system" focuses on the two unique features of such systems: the interesting behavior arises because many degrees of freedom are *interacting;* and the dynamics of the system must be *dominated* by the mutual interaction between these degrees of freedom, rather than by the intrinsic dynamics of the individual degrees of freedom. To be specific, consider the difference between the behavior of the sandpile and the ricepile. The sandpile seems to evolve into a temporal periodic state whereas the ricepile (at least for some types of rice grains) has a broad distribution of avalanche sizes. In a sandpile, the gravitational pull on individual grains dominates the friction forces between the grains – hence the tendency toward system-spanning landslides. In a ricepile, the motion of a single grain falling in the gravitational field is easily stopped by intergrain friction forces. In both systems the grains can be arranged in a huge number of different static configurations. These metastable states exist as a consequence of the finite threshold produced by the intergrain friction. However, in a sandpile the kinetic energy of the falling grains readily dominates these friction forces. In the ricepile, the friction is relatively stronger, and the effect of the toppling rice grains is to move the system between the different metastable static configurations.

One effect of the *threshold* is to allow a large number of static metastable configurations. Cafiero et al. (1995) have introduced the instructive term *local rigidity* to describe the local stabilizing effect of the threshold. Another effect of the threshold is related to the question of how criticality is reached. From the model studies in Section 5.5 it seems that the threshold is essential for the self-organization to criticality. The renormalization group analysis of sandpile-like models found an attractive fixed point. This means that the long-distance long-time behavior of these models exhibits scale invariance without any parametric fine tuning. The fixed point of the forest fire model, in contrast, was found to be repulsive. This means that the model is characterized by a specific scale. The dynamical evolution of the model does not in itself bring the model to the critical point. Of course, the forest fire and the sandpile models also differ in respects other than the presence of a threshold. Nevertheless, we shall conjecture† that the existence of local thresholds is a necessary (although certainly not sufficient) condition for self-organization to criticality.

We can expect systems to be interaction-dominated only in the limit of *slow drive*. Strong drive will not allow the system to relax from one metastable configuration to the other. The slow drive is needed in order for the intrinsic properties of the system to control the dynamics. If we pull the spring–block system of the Burridge–Knopoff model at high speed over the surface, there will be no time for the springs to relax to the equilibrium configurations; the behavior of the system will be completely dominated by the external applied drive. The slow drive is accordingly required for two reasons: first, in a sense similar to the usual weak drive encountered in linear response theory. Second, slow and therefore weak drive is necessary in order to maintain the effect of the thresholds.

6.2 What Is Tuning?

To what extent is tuning needed? And, if it *is* needed, does it then make any sense to speak of *self*-organization? I think that some degree of tuning is inevitable. Consider the OFC earthquake model (see Section 4.3). From numerical and analytical studies we believe that the model exhibits SOC only when an open boundary condition is used. In the case of periodic boundaries, oscillatory behavior occurs. One could, of course, consider this necessary restriction

† This viewpoint was already implicit in the original BTW (1987) paper and has been made more explicit by Cafiero et al. (1995).

on the choice of boundary condition as a form of "tuning." Similarly, it seems to be necessary to drive the open system homogeneously in the case of nonconservative updates ($\alpha < 1/q_c$) in order to obtain power laws. This also can certainly be considered as tuning – namely, the fluctuations in the driving term are tuned to zero. Does this then imply that SOC behavior is atypical in the same sense that critical phenomena are atypical in equilibrium statistical mechanics? Is SOC something that is encountered only under very special conditions? The question about tuning is difficult because it is concerned with how we break the world down into subsystems and their surroundings. Take the question of periodic versus open boundary conditions. Real systems are always finite, so open boudaries are likely to be more relevant than periodic ones. But then the question arises of what kind of open boundaries should be used. Of course, the answer can be given only in terms of what is relevant to the particular phenomenon under consideration. The same can be said about the way the model is driven. Whether one should use homogeneous or fluctuating drive can be answered only in the context of the particular system under examination.

It is certainly important to know the degree of self-organization or the robustness of the critical behavior of models. Self-organization to criticality will definitely occur only under certain conditions; one will always be able to generalize a model sufficiently to lose the critical behavior. Hence the question becomes just what *is* relevant in a given context. This is where the super-general approach must be supplemented by insight from the specific science to which a given system belongs. Think again of equilibrium critical points. If we examine a certain system on its own then the critical points might be interesting from a theoretical point of view, especially since so much beautiful theory can be done concerning the exponents that characterize behavior at or near the critical points. However, if we ask about typical behavior within a given temperature range, it seems as if the few isolated critical points in the phase diagram of a system are rather irrelevant. But this is a very superficial consideration. In cell biology, for example, it is found that membranes typically operate in the vicinity of their critical points, probably in order to take advantage of the increase in susceptibility close to a critical point. That is, the interactions of biological systems tune the parameters (temperature, pressure, concentration, etc.) toward the critical regimes.

This is all we want to say about question (1) in our list. Let us now turn to the less difficult questions (2)–(4). Although there is not yet a complete and well-established formalism for SOC that is similar to, say, the canonical ensemble theory of equilibrium systems, we did demonstrate in Chapter 5 that a body of formalisms has been developed. The search for the equivalent of the free energy or the partition function is, however, still ongoing. Yet with the development of exact solutions and the successful application of renormalization group techniques, SOC has been given a substance that is difficult to achieve from computer simulations alone.

Question (3) asks if we have really learned anything new from SOC. I think the most important lesson is that, in a great variety of systems, it is misleading to neglect the fluctuations. This is true in particular for slowly driven interaction-dominated threshold systems. Self-organized criticality has inspired interest in threshold dynamics and in the resulting strongly fluctuating avalanche-like temporal evolution. A smooth gradual development is, in these systems, replaced by periods of calm quiescence interrupted by hectic activity. The fluctutions are so large that the fate of a major part of the system can be determined by a single burst of activity. Dinosaurs may have become extinct simply as a result of an intrinsic fluctuation in a system consisting of a highly interconnected and interacting web of species; there may be no need for an explanation in terms of external bombardment by meteorites. Fluctuations in these systems are so large that the atypical become the typical. Indeed, complex behavior was appreciated long before the concept of SOC was introduced. Still, SOC has for the first time focused on a hitherto overlooked mechanism responsible for at least some types of complex behavior – namely, what we have labeled as SDIDT systems.

Let us finally turn to question (4), which concerns the predictive power of SOC. It is obviously not possible – given only a set of initial conditions – to predict the evolution of a SOC system in detail. Fluctuations are too important. As mentioned previously, we are not yet able to list a comprehensive set of necessary and sufficient conditions under which SOC will emerge. Notwithstanding this deficiency, we have also discussed certain features (such as SDIDT) that seem to induce a tendency of self-organization toward a crititical state. For instance, the avalanche dynamics observed in ricepiles and in superconductors (see Sections 3.3 and 3.4) are examples in which behavior predicted by SOC is observed.

Self-organized criticality has been established as a subfield of nonequilibrium statistical mechanics. At the moment, it is perhaps as much characterized by its phenomenological emphasis as by a constructive and exclusive definition. More important is that SOC has forced us to recognize that thresholds, metastability, and large fluctuations play a key role in the spatiotemporal behavior of a large class of many-body systems. This new insight is sufficiently important to justify and inspire more theoretical, observational, and experimental research from the vantage point of self-organized criticality.

A

Code for the BTW Sandpile

We list here one example of how the updating algorithm (see (4.9)) of the BTW model can be simulated in two dimensions. The implementation given is meant to be simple and easy to understand. This piece of code is included as inspiration to people who have little or no experience in programming this kind of problem. The code is written in FORTRAN simply because the author knows how to program in this language. A change to any other language should not cause problems. The code given is not fast; it is, in fact, very slow. But as soon as one gets the hang of it, optimizing a code by gradual improvements is reasonably straightforward. For a particularly fast and efficient implementation of BTW-like models, see Grassberger and Manna (1990), Manna (1990), and Grassberger (1994).

```
c     The basic dynamical variable z(r) is
c     stored in the two-dimensional integer array

      Integer z(1:L,1:L)

c     First we define an initial random configuration.
c     Here ranf(Nseed) is a random number generator that returns
c     a random (uniformly distributed) number in the interval [0, 1).
c     One should be careful to make sure to use a random number
c     generator of high quality. All z-values will be integers
c     between zero and Zc.

      do 10 j=1,L
      do 10 i=1,L
           z(i,j) = Zc*ranf(Nseed)
      10 continue

c     Next we must add a grain of sand and check if we make
c     the sandpile overcritical by the extra grain added.

29    continue
      iadd = L*ranf(Nseed) + 1
      jadd = L*ranf(Nseed) + 1
```

131

```
      z(iadd,jadd) = z(iadd,jadd) + 1
      if ( z(iadd,jadd) .gt. Zc) then
            call relax(z,Zc)
            goto 29
      else
            goto 29
      end if

      stop
      end
```

c   The next subroutine performs the relaxation of the
c   lattice. We need to store the new values of the z-variables
c   during the update. This is because the whole lattice is
c   simultaneously updated. We store the intermediate values in
c   the array z0(i,j).

```
      Subroutine relax(z,Zc)

      Integer z(1:L,1:L), z0(1:L,1:L)
```

c   We first store the current values of the z-array.

```
      do 10 j=1,L
      do 10 i=1,L
            z0(i,j) = z(i,j)
10    continue

5     continue
```

c   Then we go through the entire lattice and perform the
c   update where needed. The array z0(i,j) is successively
c   updated in order to accumulate the effect of the updates.

```
      do 20 j=1,L
      do 20 i=1,L
            if(z(i,j).le.Zc)then
            goto 20
      else
            z0(i,j)   = z0(i,j)   - 4
            z0(i-1,j) = z0(i-1,j) + 1
            z0(i+1,j) = z0(i+1,j) + 1
            z0(i,j-1) = z0(i,j-1) + 1
            z0(i,j+1) = z0(i,j+1) + 1
      end if
20    continue
```

```
c       We can now store the total effect of a simultaneous
c       update of the entire lattice. While we store the new
c       z-values we also check if more relaxation is needed.
c       This is the use of the variable "flag".

        flag = 0
        do 30 j=1,L
        do 30 i=1,L
            z(i,j) = z0(i,j)
            if(z0 .gt. Zc) flag = 1
30      continue

        if (flag .eq. 1) then

c       More relaxation needed.

            goto 5
        else

c       No more overcritical sites. We can return
c       to the main program.

            return
        end if

        end
```

The code sketched here can be speeded up significantly by two simple improvements. If one adds only a single grain of sand at a time then we know that, if an avalanche is induced, it will be induced at the position where the new grain was added. Hence we need not search the entire lattice for overcritical sites. Successive overcritical sites will always occur in the neighborhood of previous overcritical sites. The subroutine relax(z,Zc) can accordingly be speeded up by checking only during update number $t + 1$ the neighbor sites of the sites that were overcritical in update number t. Another improvement is to change the arrays z(1:L,1:L) and z0(1:L,1:L) into vectors z(1:L^2) and z0(1:L^2). This is done by enumerating the sites on the square lattice by number from 1 to L^2. One must now keep in mind that the neighbors of site number k are the sites $k - 1$, $k + 1$, $k - L$, and $k + L$.

The boundary sites need special attention. The code given here must be altered in this respect. When updating the lattice in subroutine relax(z,Zc), one needs to separate the updating of bulk sites from edge sites. The edge sites will then have to be updated according to one of the boundary conditions discussed in Section 4.2.2.

B

Code for the Lattice Gas

We discuss here some details of writing a computer code for the dynamics of the lattice gas defined in Section 4.4. The code is not optimized; it is included as a guide and for inspiration only.

The configuration of the particles on the (two-dimensional) lattice is stored in the array n(1:L,1:L). The elements of this array take the value n(i,j)=1 if the site (i,j) is occupied by a particle; otherwise, n(i,j)=0. The update, as defined in Section 4.4.1, can be coded as sketched below. A complete code of the lattice gas will include an initiation of the array n(i,j), and special attention must be paid to the boundary. Since the lattice is updated simultaneously, we need the array nnew(1:L,1:L) to store intermediate changes in the particle configuration. We update the lattice by visiting all the *empty* sites of the lattice. An empty site (i,j) becomes an occupied site if a particle on one of the eight neighbor sites is able to move onto site (i,j). We therefore visit all eight neighbor sites to calculate the forces acting on particles in these sites. We then check if one of these particles has a force pointing toward site (i,j) and is of a magnitude greater than any other force pointing toward site (i,j).

```
      Integer n(1:L,1:L), nnew(1:L,1:L)

c     Update of the lattice configuration

      do 10 j=1,L
      do 10 i=1,L
      if (n(i,j) .eq.1) then
           goto 10
      else
           call forces(i,j,n,nnew)
      end if
10    continue

c     The following subroutine performs the update. The forces
c     on the particles located in the neighbor sites of the empty
c     site (i,j) are calculated and their magnitude stored in the
```

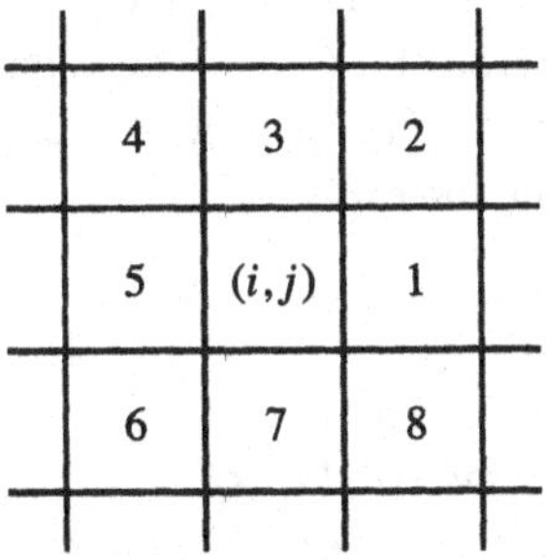

Figure B.1. Enumeration of the eight neighbor sites surrounding the site (i,j).

```
c     vector absforce(1:8). The array neigh(1:2,1:8) should
c     be defined in the beginning of the main program. The array
c     allows one to access the nearest neighbor and the next nearest
c     neighbor sites of a given site by looping through a do-loop.
c     The array is defined in the following way (see also Figure B.1).

      neigh(1,1) = 1
      neigh(2,1) = 0
      neigh(1,2) = 1
      neigh(2,2) = 1
      neigh(1,3) = 0
      neigh(2,3) = 1
      neigh(1,4) = -1
      neigh(2,4) = 1
      neigh(1,5) = -1
      neigh(2,5) = 0
      neigh(1,6) = -1
      neigh(2,6) = -1
      neigh(1,7) = 0
      neigh(2,7) = -1
      neigh(1,8) = 1
      neigh(2,8) = -1

      subroutine forces(i,j,n,nnew)
      Integer n(1:L,1:L), nnew(1:L,1:L),
         nlabel(1:8), neigh(1:2,1:8)
      Real absforce(1:8)
      common/neighbors/neigh
```

```
c      The variable nc counts the number of occupied neighbor
c      sites. The variable nlabel(1:8) stores the label of the
c      occupied neighbor site. The variable absforce(1:8) stores
c      the magnitude of the forces.
       nc = 0
       do 10 k = 1,8
            in = i + neigh(1,k)
            jn = j + neigh(2,k)
            if ( n(in,jn).eq.0) then
                goto 10
            else
                Fx = n(in+1,jn)-n(in-1,jn)
                Fy = n(in,jn+1)-n(in,jn-1)
                ff = sqrt(Fx**2 +Fy**2)
c      We now check if the normalized integer force points toward the
c      empty site (i,j).
                nx = nint(Fx/ff)
                ny = nint(Fy/ff)
                if( nx.eq.-neigh(1,k) .and.
                    ny.eq.-neigh(2,k)) then
                    nc = nc + 1
                    nlabel(nc) = k
                absforce(nc) = ff
                else
                    goto 10
                end if
            end if
10     continue
c      In case more than one of the neighbor particles feels a force
c      pointing toward the site (i,j), we need to decide which one
c      of these, if any, should be moved onto the site (i,j). To make
c      this decision we need to compare the amplitude of the forces.
       if (nc.le.1) then
            goto 40
       else
c      The variable fmax is used to find the maximum force
            fmax = -10.0
            do 20 m = 1,nc
```

```
          if (absforce(m) .ge. fmax) then
              fmax = absforce(m)
              mmax = m
          end if
20        continue
```

c   We now check how many particles are subject
c   to the maximum force.

```
     ncmax = 0
     do 30 m = 1,nc
          if (absforce(m) .eq. fmax) then
              ncmax = ncmax + 1
          else
              goto 30
          end if
30   continue

          if(ncmax .gt. 1) then
```

c   No particle is allowed to move onto the empty site (i,j).

```
              return
          else
              goto 40
          end if
```

```
40   continue
```

c   The appropriate particle is moved onto site (i,j).

```
     nnew(i,j) = 1
     k = nlabel(mmax)
     ii = i + neigh(1,k)
     jj = j + neigh(2,k)
     nnew(ii,jj) = 0
```

```
     return
```

As mentioned in Appendix A, one can speed up the code by representing the array $n(1:L,1:L)$ by a vector $n(1:L^2)$. More importantly, note that the code presented here calculates the forces on a given particle more than once whenever that particle is neighbor to more than one empty site. This is obviously a waste of CPU time. An improved code would first calculate and store all forces on particles neighbor to empty sites. These forces would then simply be recalled when needed.

C

Code for the Bak–Sneppen Evolution Model

In this appendix we give some hints concerning the coding of the M-update version of the Bak–Sneppen model (see Section 4.6.2).

The barriers are stored in the vector $B(1:L)$. The probability density $P(B)$ of the barriers is stored in $prob(1:100)$, and the probability density p_{min} for the minimal barrier is stored in $probm(1:100)$. The following performs the update according to the algorithm in Section 4.6.2. Statistics for $P(B)$ and p_{min} are also accumulated. As emphasized previously, one requires a high-quality random number generator, which we denote by $ranf(Nseed)$.

```
      subroutine update
      double precision B(1:L),
         prob(0:100),probm(0:100)
      double precision ranf

c     The following integer arrays are used when the M
c     smallest barriers are located.

      integer Mmin(1:L),Mmin0(1:L),new(1:L)

c     The statistical arrays are cleared.
      do 5 nn = 1,100
            prob(nn) = 0
            probm(nn) = 0
5     continue

c     The M-minimal B-values are localized.
c     *****************************************************

      do 10 k =1,L

c     Whenever a barrier has been found to be a minimal
c     barrier, it is excluded from the search. The array new(k)
c     stores the address of the barriers remaining to be searched.

            new(k) = k
```

```
c       We accumulate the number of times the barrier value
c       B(k) has occurred.

        nn = B(k)*100
        prob(nn) = prob(nn) + 1
10 continue

c       We now start to identify the M smallest barriers.

     do 20 i = 1,M
        Bmin = 10.
        do 30 n = 1,L+1-i
           nn = new(n)
           if(B(nn).le.Bmin)then
              Bmin = B(nn)
              Mmin(i) = nn
           endif
30       continue
        ncount = 1
        do 40 j =1,L-1
           if(new(j).lt.Mmin(i))goto 40
           new(j) = new(j+1)
40       continue
20   continue

c       The smallest minimal-value:

     Bmin = 10.
     do 45 i = 1,M
        if(B(Mmin(i)).le.Bmin) Bmin = B(Mmin(i))
45   continue
     nn = Bmin*100.
     probm(nn) = probm(nn) + 1

c       The M-minimal B-value sites and their neighbors are updated.
c       ******************************************************

     do 50 n0 = 1,M
        n = Mmin(n0)
        nl = n - 1
        nr = n + 1

c       The next two lines take care of the periodic boundary condition.
```

```
      if(n.eq.1) nl = L
      if(n.eq.L) nr = 1
      B(nl) = ranf(Nseed)
      B(n)  = ranf(Nseed)
      B(nr) = ranf(Nseed)
50    continue

return
end
```

It is easy to add to this code subroutines that calculate the distribution of, say, distances between updated sites, durations of avalanches (see Section 4.6.2 for the definition of an avalanche in this model), the two-point correlation function, and/or the power spectrum of the sequence B(1), B(2), ..., B(L).

The code can be made many orders of magnitude faster by dividing the interval [0, 1] into a set of Q subintervals $I_1 = [0, 1/Q]$, $I_2 = [1/Q, 2/Q]$, ... The search for the M smallest barriers can then be restricted to those few sites with barriers in the interval I_1, or in I_2 if I_1 contains fewer than M barriers. Employing a "link list" makes this search and bookkeeping very fast.

D

Power Spectra and the Correlation Function

To get a feeling for the autocorrelation function $G(\tau)$ defined in (2.1), let us consider a signal that can assume only (and with equal probability) the two values F and $-F$. The signal switches at random from $N(\tau) = F$ to $N(\tau) = -F$ with a constant probability λ per time unit (MacDonald 1962). Thus, $\langle N(\tau_0)\rangle_{\tau_0} = 0$. The autocorrelation function can be calculated according to the number of times the signal has managed to switch during the time interval τ:

$$G(\tau) = FP(N(\tau_0) = F)\{Fp_0(\tau) - Fp_1(\tau) + Fp_2(\tau) - \cdots\}$$
$$+ (-F)P(N(\tau_0) = -F)\{-Fp_0(\tau) + Fp_1(\tau) - Fp_2(\tau) + \cdots\}. \tag{D.1}$$

The first line in this formula describes the case where the signal is up $N(\tau_0) = F$ at time τ_0. This happens with probability $P(N(\tau_0) = F) = 1/2$. The value of the signal at τ units later depends on how many times the signal has switched. No switching occurs with probability $p_0(\tau)$, in which case one obviously still has $N(\tau_0 + \tau) = F$. Precisely one switch occurs with probability $p_1(\tau)$, and so forth. The second line in (D.1) describes the same scenario but with $N(\tau_0) = -F$. The equation reduces to

$$G(\tau) = F^2\{p_0(\tau) - p_1(\tau) + p_2(\tau) - \cdots\}. \tag{D.2}$$

The switching probabilities are easily calculated by standard arguments for Poisson processes. One derives a differential equation for $p_0(\tau)$ in the following way. The probability $p_0(\tau + d\tau)$ that no switching occurs during the time interval $\tau + d\tau$ is equal to the probability $p_0(\tau)$ that no switching occurs during the interval τ times the probability $1 - \lambda\,d\tau$ that no switch takes place during the infinitesimal interval $d\tau$. We have, accordingly,

$$p_0(\tau + d\tau) = p_0(\tau)[1 - \lambda\,d\tau]. \tag{D.3}$$

By expanding the left-hand side,

$$p_0(\tau + d\tau) \simeq p_0(\tau) + d\tau\frac{dp_0(\tau)}{d\tau},$$

we arrive at the formula

$$\frac{dp_0}{d\tau} = -\lambda p_0. \tag{D.4}$$

This equation uniquely determines p_0 to be $p_0(\tau) = \exp(-\lambda\tau)$ when we make use of the fact that $p_0(0) = 1$. No switching occurs with probability 1 during an interval of time $\tau \to 0$. We can readily calculate $p_n(\tau)$ when we know $p_0 = (\tau)$. The probability that precisely one switch occurs at some instant τ_1 is equal to the probability that no switch $p_0(\tau_1)$ occurs during the time τ_1, times the probability $\lambda\, d\tau_1$ that the switch occurs during the time interval $d\tau_1$, times the probability $p_0(\tau - \tau_1)$ that no further switch occurs during the interval τ. That is,

$$p_1(\tau) = \int_0^\tau e^{-\lambda\tau_1}\lambda\, d\tau_1 e^{-\lambda(\tau-\tau_1)}$$
$$= \tau\lambda e^{-\lambda\tau}. \tag{D.5}$$

The probability for precisely two switches is

$$p_2(\tau) = \int_0^\tau p_0(\tau_1)\lambda\, d\tau_1 \int_{\tau_1}^\tau p_0(\tau_2 - \tau_1)\lambda\, d\tau_2\, p_0(\tau - \tau_2)$$
$$= \frac{(2\lambda\tau)^2}{2}e^{-\lambda\tau}. \tag{D.6}$$

The general expression is

$$p_n(\tau) = \frac{(\lambda\tau)^n}{n!}e^{-\lambda\tau}. \tag{D.7}$$

From (D.2) we obtain

$$G(\tau) = F^2 e^{-\lambda\tau}\left\{1 - \lambda\tau + \frac{(\lambda\tau)^2}{2!} - \cdots\right\}$$
$$= F^2 \exp(-2\lambda\tau)$$
$$= F^2 \exp(-2\lambda|\tau|). \tag{D.8}$$

The last equality follows because the autocorrelation function is an even function. We conclude from (D.8) that correlations in the random switching process decay exponentially over a time scale given by $1/\lambda$, the switching rate

$$S(f) = \lim_{T \to \infty} \frac{1}{2T} \left| \int_{-T}^{T} d\tau \, N(\tau) e^{2\pi i f \tau} \right|^2$$

$$= \lim_{T \to \infty} \frac{1}{2T} \int_{-T}^{T} d\tau_1 \int_{-T}^{T} d\tau_2 \, N(\tau_1) N(\tau_2) e^{2\pi i f (\tau_2 - \tau_1)}$$

$$= \lim_{T \to \infty} \int d(\tau_2 - \tau_1) e^{2\pi i f (\tau_2 - \tau_1)}$$

$$\times \frac{1}{2T} \int d\tau_1 \, N(\tau_1) N(\tau_1 + (\tau_2 - \tau_1)). \tag{D.9}$$

We have deliberately omitted the limits of the integrals in the last equation in order not to be swamped by technical details. The step consists of making use of the fact that

$$\lim_{T \to \infty} \frac{1}{2T} \int_{-T}^{T} d\tau_0 \, N(\tau_0) N(\tau_0 + \tau) = G(\tau) + \langle N(\tau_0) \rangle^2_{\tau_0}; \tag{D.10}$$

see (2.1). The second term in this equation is independent of time and will therefore contribute only to the $f = 0$ value of $S(f)$. We neglect this term. Finally, since the autocorrelation function $G(\tau)$ is an even function, the Fourier transformation reduces to a cosine transformation

$$S(f) = \lim_{T \to \infty} 2 \int_0^T G(\tau) \cos(2\pi f \tau) \, d\tau. \tag{D.11}$$

It is clear from (D.11) why a power spectrum of the form $S(f) \sim 1/f^\beta$ with $\beta \simeq 1$ is of special interest (see Section 2.3).

Let us sketch how we can obtain a $1/f$ spectrum from a linear superposition of signals each having exponentially decaying correlation functions (van der Ziegle 1950). Consider a signal for which we know that the correlation function has the form

$$G(\tau_0) = A e^{-\tau/\tau_0}. \tag{D.12}$$

The power spectrum of this signal is, according to (D.11), given by

$$S_{\tau_0}(f) = \frac{4A\tau_0}{1 + (2\pi f \tau_0)^2}. \tag{D.13}$$

Assume now that a set of these τ_0 signals are operating simultaneously without interference or correlations between different τ_0 signals. The probability that one of the signals in operation is characterized by the correlation time τ_0 is given by $P(\tau_0)$. Since we assume that there are no correlations between the individual τ_0 signals, the power spectrum of the total signal is simply given by the weighted sum of the power spectrum of the individual signals:

$$S(f) = \int_0^\infty d\tau_0 \, \frac{4A\tau_0}{1 + (2\pi f \tau_0)^2} P(\tau_0). \tag{D.14}$$

Assume, furthermore, that the probability density of the correlation times is of the form

$$P(\tau_0) = \begin{cases} K/\tau_0 & \text{if } \tau_0 \in [\tau_1, \tau_2], \\ 0 & \text{otherwise.} \end{cases} \tag{D.15}$$

From (D.14) it follows immediately that

$$S(f) = \frac{2AK}{\pi f}(\arctan(2\pi\tau_2 f) - \arctan(2\pi\tau_1 f)). \tag{D.16}$$

If $\tau_1 \ll 1/f \ll \tau_2$ we have $\arctan(2\pi\tau_2 f) \simeq \pi/2$ and $\arctan(2\pi\tau_1 f) \simeq 0$. In this way we obtain

$$S(f) \simeq \frac{2AK}{\pi f} \quad \text{for} \quad \frac{1}{2\pi\tau_2} \ll f \ll \frac{1}{2\pi\tau_1}. \tag{D.17}$$

The problem with deriving the $1/f$ spectrum in this way is that one must find a convincing argument for why the rather ad hoc form of the density of the correlation times in (D.15) is particularly widespread in nature.

The idea that the $1/f$ spectrum can arise from a linear superposition of independent signals has also been used in the discussion of some SOC models (see e.g. the remarks in Section 4.2.5). Consider the total flow down the slope of a sandpile. We construct this signal by adding together individual avalanche signals commenced at a set $\mathcal{T} = \{\ldots, \tau_{-2}, \tau_{-1}, \tau_0, \tau_1, \tau_2, \ldots\}$ of random starting times

$$J(\tau) = \sum_{\tau_i < \tau} f_{A_i}(\tau - \tau_i). \tag{D.18}$$

It is easy to show that the autocorrelation function of $J(\tau)$ is given by the weighted average of the autocorrelation functions of the individual signals in the following sense:

$$G_J(\tau) = \nu \sum_A P(A) \int_0^\infty dt \, f_A(t) f_A(t + |\tau|). \tag{D.19}$$

Here ν denotes the rate per time with which the individual avalanche signals occur in (D.18); $P(A)$ is the probability that an avalanche of type A is induced when a grain is added to the pile. Let us assume that an avalanche signal is completely determined by its total size S and its duration T.† In a rough approximation, we can represent the time signal of an avalanche by a "box" signal

---

† This is obviously not completely correct. There are many different functions $\tau \to f_A(\tau)$ with support restricted to $\tau \in [0, T]$ and integral $S = \int_0^T d\tau \, f_A(\tau)$. It turns out, however, that the avalanche signals of the BTW model are sufficiently determined by S and T. Many details of the discussion presented here can be found in Jensen et al. (1989), Kertész and Kiss (1990), and Christensen et al. (1991).

$$f_A(t) = f_{S,T}(t) = \begin{cases} S/T & \text{if } t \in [0, T], \\ 0 & \text{otherwise.} \end{cases} \tag{D.20}$$

From (D.19) we then obtain

$$G(\tau) = \nu \int_{|\tau|}^{\infty} dT \int_{0}^{\infty} dS\, P(S, T)\left(\frac{S}{T}\right)^2 (T - |\tau|). \tag{D.21}$$

We substitute (D.21) into (D.11), perform two partial integrations of (D.11), and make use of the lifetime distribution function introduced in (4.26) to derive (4.25).

E

Statistical Weights in the DDRG

There are four different sets of configurations (see Figure E.1). The weight factor W_α can be written as a product of the probability that α sites are occupied (and $4 - \alpha$ are unoccupied) by a degeneracy factor n_α: $W_\alpha = n_\alpha \rho^\alpha (1 - \rho)^{4-\alpha}$. The expressions for n_α are given in Figure E.1.

The calculations are performed as follows. Let us first consider the $\alpha = 2$ case. There are $K_{4,2} = 6$ configurations for which two of the four sites are critical. Two of these configurations correspond to the critical sites being placed along the diagonal. Such configurations are ignored because they do not meet the spanning criterion. This leaves us with four configurations. For each of these remaining configurations, we have two possible choices when we turn one of the critical sites into an unstable site. In total there are eight different but dynamically equivalent configurations of $\alpha = 2$. Next, we turn our attention to $\alpha = 3$ configurations. We can pick three critical sites in $K_{4,3} = 4$ different

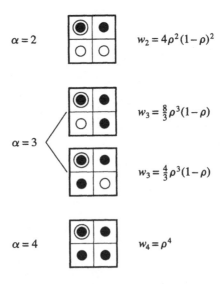

$$\alpha = 2 \qquad w_2 = 4\rho^2(1-\rho)^2$$

$$\alpha = 3 \qquad w_3 = \tfrac{8}{3}\rho^3(1-\rho)$$

$$\qquad\qquad w_3 = \tfrac{4}{3}\rho^3(1-\rho)$$

$$\alpha = 4 \qquad w_4 = \rho^4$$

Figure E.1. The α-configurations and their statistical weights.

ways. The $\alpha = 3$ sites break up into two classes of different dynamical behavior depending on where we choose to position the unstable site among the three critical sites. There are four configurations with the unstable site positioned between the two critical sites (see Figure E.1) and eight configurations where the critical site is located at either end of the string of three sites. Finally, there is only one configuration for which $\alpha = 4$, but the unstable site can be chosen in four different ways. The probability that a critical site in a given α-configuration is hit by a perturbation (and thereby turned into an unstable site) is equal to $1/\alpha$. Accordingly, the n_α factors are given by

$$n_2 = 4, \quad n_3^a = \tfrac{4}{3}, \quad n_3^b = \tfrac{8}{3}, \quad n_4 = 1. \tag{E.1}$$

References

Andersen, J. V., Jensen, H. J., and Mouritsen, O. G. (1991), *Phys. Rev. B* 44: 429.

Bak, P. (1991), in T. Riste and D. Sherrington (eds.), *Spontaneous Formation of Space-Time Structures and Criticality*. Dordrecht: Kluwer.

Bak, P. (1996), *How Nature Works*. New York: Springer-Verlag.

Bak, P., and Sneppen, K. (1993), *Phys. Rev. Lett.* 74: 4083.

Bak, P., Tang, C., and Wiesenfeld, K. (1987), *Phys. Rev. Lett.* 59: 381.

Bak, P., Tang, C., and Wiesenfeld, K. (1988), *Phys. Rev. A* 38: 364.

Barabási, A.-L., and Stanley, H. E. (1995), *Fractal Concepts in Surface Growth*. Cambridge University Press.

Binney, J. J., Dowrick, N. J., Fisher, A. J., and Newman, M. E. J. (1992), *The Theory of Critical Phenomena*. Oxford, UK: Oxford University Press.

Bonabeau, E., and Lederer, P. (1994), *J. Phys. A: Math. Gen.* 27: L243.

Boyce, W. G., and Di Prima, R. C. (1965), *Elementary Differential Equations and Boundary Value Problems*. New York: Wiley, chap. 9.

Bretz, M., Cunningham, J. B., Kurczynski, P. L., and Nori, F. (1992), *Phys. Rev. Lett.* 69: 2431.

Burridge, R., and Knopoff, L. (1967), *Bull. Seismol. Soc. Am.* 57: 341.

Cafiero, R., Loreto, V., Pietronero, L., Vespignani, A., and Zapperi, S. (1995), *Europhys. Lett.* 29: 111.

Ceva, H. (1995), *Phys. Rev. E* 52: 154.

Christensen, K. (1992), "Self-Organization in Models of Sandpiles, Earthquakes, and Flashing Fireflies," Ph.D. Thesis, Department of Physics, University of Århus, Denmark.

Christensen, K., Flyvbjerg, H., and Olami, Z. (1993), *Phys. Rev. Lett.* 71: 2737.

Christensen, K., Fogedby, H. C., and Jensen, H. J. (1991), *J. Stat. Phys.* 63: 653.

Christensen, K., and Olami, Z. (1992a), *Phys. Rev. A* 46: 1829.

Christensen, K., and Olami, Z. (1992b), *J. Geophys. Res.* 97: 8729.

Christensen, K., and Olami, Z. (1993), *Phys. Rev. E* 48: 3361.

Clar, S., Drossel, B., and Schwabl, F. (1994), *Phys. Rev. E* 50: 1009.

Corral, A., and Diaz-Guilera, A. (1997), *Phys. Rev. E* (to appear).

Corral, A., Perez, C. J., Diaz-Guilera, A., and Arenas, A. (1995), *Phys. Rev. Lett.* 74: 118.

Datta, A., Gilhøj, H., and Jensen, H. J. (1997), in preparation.

de Boer, J., Derrida, B., Flyvbjerg, H., Jackson, A. D., and Wettig, T. (1994), *Phys. Rev. Lett.* 73: 906.

Dhar, D. (1990), *Phys. Rev. Lett.* 64: 1613.

Dhar, D., and Majumdar, S. N. (1990), *J. Phys. A: Math. Gen.* 23: 4333.

Dhar, D., and Manna, S. S. (1994) *Phys. Rev. E* 49: 2684.

Dhar, D., and Ramaswamy, R. (1989), *Phys. Rev. Lett.* 63: 1659.

Diaz-Guilera, A. (1992), *Phys. Rev. A* 45: 8551.

Diaz-Guilera, A. (1994), *Europhys. Lett.* 26: 177.

Drossel, B., and Schwabl, F. (1992), *Phys. Rev. Lett.* 69: 1629.

Drossel, B., and Schwabl, F. (1993), *Physica A* 199: 183.

Duta, P., and Horn, P. M. (1981), *Rev. Mod. Phys.* 53: 497.

Erzan, A., Pietronero, L., and Vespignani, A. (1995), *Rev. Mod. Phys.* 67: 545.

Esipov, S. E., and Newman, T. J. (1997), in T. Poeschel and L. Schimanshy-Geier (eds.), *Lectures on Stochastic Dynamics*. (Lecture Notes in Physics). Berlin: Springer-Verlag.

Feynman, R. (1964), *The Feynman Lectures on Physics* (R. B. Leighton and M. Sands, eds.). Reading, MA: Addison-Wesley.

Field, S., Witt, J., and Nori, F. (1995), *Phys. Rev. Lett.* 74: 1206.

Fiig, T., and Jensen, H. J. (1993), *J. Stat. Phys.* 71: 653.

Flyvbjerg, H., Sneppen, K., and Bak, P. (1993), *Phys. Rev. Lett.* 71: 4087.

Fogedby, H. C., Jensen, M. H., Zhang, Y.-C., Bohr, T., Jensen, H. J., and Rugh, H. H. (1991), *Mod. Phys. Lett. B* 5: 1837.

Forster, D., Nelson, D. R., and Stephen, M. J. (1977), *Phys. Rev. A* 16: 732.

Frette, V., Christensen, K., Malthe-Sørensen, A., Feder, J., Jøssang, T., and Meakin, P. (1996), *Nature* 379: 49.

Gabrielov, A. (1993), *Physica A* 195: 253.

Gabrielov, A., Newman, W. I., and Knopoff, L. (1994), *Phys. Rev. E* 50: 188.

Ghaffari, P., and Jensen, H. J. (1996), *Europhys. Lett.* 35: 397.

Ghaffari, P., Lise, S., and Jensen, H. J. (1997), *Phys. Rev. E* (submitted).

Gould, S. J. (1977), *Paleobiology* 3: 135.

Grassberger, P. (1993), *J. Phys. A* 26: 2081.

Grassberger, P. (1994), *Phys. Rev. E* 49: 2436.

Grassberger, P., and Kantz, H. (1991), *J. Stat. Phys.* 63: 685.

Grassberger, P., and Mann, S. S. (1990), *J. Phys. I* (France) 51: 1077.

Grinstein, G., Hwa, T., and Jensen, H. J. (1992), *Phys. Rev. A* 45: R559.

Grinstein, G., Lee, D.-H., and Sachdev, S. (1990), *Phys. Rev. Lett.* 64: 1927.

Harris, T. E. (1963), *The Theory of Branching Processes*. Berlin: Springer-Verlag.

Held, G. A., Solina II, D. H., Keane, D. T., Haag, W. J., Horn, P. M., and Grinstein, G. (1990), *Phys. Rev. Lett.* 65: 1120.

Jaeger, H. M., Liu, C.-H., and Nagel, S. R. (1989), *Phys. Rev. Lett.* 62: 40.

Jaeger, H. M., and Nagel, S. R. (1992), *Science* 255: 1523.

Jánosi, I. M., and Horváth, V. K. (1989), *Phys. Rev. A* 40: 5232.

Jánosi, I. M., and Kertész, J. (1993), *Physica* (Amsterdam) 220A: 179.

Jensen, H. J. (1990), *Phys. Rev. Lett.* 64: 3103.

Jensen, H. J. (1991), *Physica Scripta* 43: 593.

Jensen, H. J. (1995), *J. Phys. A: Math. Gen.* 28: 1861.

Jensen, H. J., Christensen, K., and Fogedby, H. C. (1989), *Phys. Rev. B* 40: 7425.

Kadanoff, L. P. (1976), "Scaling, Universality and Operator Algebra," in C. Domb and M. S. Green (eds.), *Phase Transitions and Critical Phenomena*, vol. 5a. New York: Academic Press.

Kadanoff, L. P., Nagel, S. R., Wu, L., and Zhou, S. (1989), *Phys. Rev. A* 39: 6524.

Kardar, M., Parisi, G., and Zhang, Y.-C. (1986), *Phys. Rev. Lett.* 56: 889.

Kertész, J., and Kiss, L. B. (1990), *J. Phys. A* 23: L433.

Lee, J., and Kosterlitz, J. M. (1991), *Phys. Rev. B* 43: 3265.

Leung, K.-T., Müller, J., and Andersen, J. V. (1997), *J. Phys. I* (France) 7: 423.

Lise, S., and Jensen, H. J. (1995), unpublished manuscript.

Lise, S., and Jensen, H. J. (1996), *Phys. Rev. Lett.* 76: 2326.

Loreto, V., Pietronero, L., Vespignani, A., and Zapperi, S. (1995), *Phys. Rev. Lett.* 66: 465.

Loreto, V., Vespignani, A., and Zapperi, S. (1996), *J. Phys. A* 29: 2981.
MacDonald, D. K. C. (1962), *Noise and Fluctuations: An Introduction.* New York: Wiley.
Majumdar, S. N., and Dhar, D. (1991), *J. Phys. A: Math. Gen.* 24: L357.
Manna, S. S. (1990), *J. Stat. Phys.* 59: 509.
Manna, S. S. (1991a), *Physica* (Amsterdam) 179A: 249.
Manna, S. S. (1991b), *J. Phys. A* 24: L363.
Manna, S. S., Kiss, L. B., and Kertész, J. (1990), *J. Stat. Phys.* 61: 923.
McNamara, B., and Wiesenfeld, K. (1990), *Phys. Rev. A* 41: 1867.
Medina, E., Hwa, T., Kardar, M., and Zhang, Y.-C. (1989), *Phys. Rev. A* 39: 3053.
Middleton, A. A., and Tang, C. (1995), *Phys. Rev. Lett.* 74: 742.
Morse, P. M., and Feshbach, H. (1953), *Methods of Theoretical Physics.* New York: McGraw-Hill.
Mossner, W., Drossel, B., and Schwabl, F. (1992), *Physica A* 190: 205.
Mousseau, N. (1996), *Phys. Rev. Lett.* 77: 968.
Nicolis, G. (1989), in P. Davids (ed.), *The New Physics.* Cambridge University Press.
O'Brien, K. P., and Weissman, M. B. (1992), *Phys Rev. A* 46: R4475.
O'Brien, K. P., and Weissman, M. B. (1994), *Phys Rev. E* 50: 3446.
Olami, Z., and Christensen, K. (1992), *Phys. Rev. A* 46: R1720.
Olami, Z., Feder, H. J. S., and Christensen, K. (1992), *Phys. Rev. Lett.* 68: 1244.
Omori, F. (1895), *J. College Sci., Imper. Univ. Tokyo* 7: 111.
Paczuski, M., Maslov, S., and Bak, P. (1994), *Europhys. Lett.* 28: 295.
Paczuski, M., Maslov, S., and Bak, P. (1996), *Phys. Rev. E* 53: 414.
Percović, O., Dahmen, K., and Sethna, J. P. (1995), *Phys. Rev. Lett.* 75: 4528.
Pietronero, L., and Schneider, W. R. (1991), *Phys Rev. Lett.* 66: 2336.
Pietronero, L., Vespignani, A., and Zapperi, S. (1994), *Phys. Rev. Lett.* 72: 1690.
Pla, O., Wilkin, N. K., and Jensen, H. J. (1996), *Europhys. Lett.* 33: 297.
Plourde, B., Nori, F., and Bretz, M. (1993), *Phys. Rev. Lett.* 71: 2749.
Press, W. H. (1978), *Comments Astrophys. Space Phys.* 7: 103.
Priezzhev, V. B., Ktitarev, D. V., and Ivashkevich, E. V. (1996), *Phys. Rev. Lett.* 76: 2093.
Raup, M. D. (1986), *Science* 231: 1528.
Reif, F. (1965), *Fundamentals of Statistical and Thermal Physics.* New York: McGraw-Hill.
Rhodes, C., and Anderson, R. (1996), *Nature* 381: 600.
Rhodes, C., Jensen, H. J., and Anderson, R. (1997), *Philos. Trans. Roy. Soc. London Ser. B* (submitted).
Rosendahl, J., Vekić, M., and Kelly, J. (1993), *Phys. Rev. E* 47: 1401.
Roux, S., and Hansen, A. (1994), *J. Physique I* (France) 4: 515.
Scholz, C. H. (1990), *The Mechanics of Earthquakes and Faulting.* Cambridge University Press.
Scholz, C. H. (1991), in T. Riste and D. Sherrington (eds.), *Spontaneous Formation of Space-Time Structures and Criticality.* Dordrecht: Kluwer.
Sethna, J. P., Dahmen, K., Kartha, S., Krumhansl, J. A., Roberts, R. W., and Shore, J. D. (1993), *Phys. Rev. Lett.* 70: 3347.
Sneppen, K. (1992), *Phys. Rev. Lett.* 69: 3539.
Sneppen, K., and Jensen, M. H. (1993), *Phys. Rev. Lett.* 71: 101.
Socolar, J., Grinstein, G., and Jayaprakash, C. (1993), *Phys. Rev. E* 47: 2366.
Sornette, D. (1991), in T. Riste and D. Sherrington (eds.), *Spontaneous Formation of Space-Time Structures and Criticality.* Dordrecht: Kluwer.

References

Sornette, D. (1994), *J. Phys. I* (France) 4: 209.
Stauffer, D., and Aharony, A. (1992), *Introduction to Percolation Theory*. London: Taylor and Francis.
Tang, C., Wiesenfeld, K., Bak, P., Coppersmith, S., and Littlewood, P. (1987), *Phys. Rev. Lett.* 58: 1161.
Tinkham, M. (1975), *Introduction to Superconductivity*. New York: McGraw-Hill.
van der Ziegle, A. (1950), *Physica* 16: 359.
Vespignani, A., Zapperi, S., and Pietronero, L. (1995), *Phys. Rev. E* 51: 1711.
Waldrop, M. M. (1992), *Complexity*. New York: Penguin.
Weissman, M. B. (1988), *Rev. Mod. Phys.* 60: 537.
Wiesenfeld, K., Tang, C., and Bak, P. (1989), *J. Stat. Phys.* 54: 1441.
Yeh, W. J., and Kao, Y. H. (1984), *Phys. Rev. Lett.* 53: 1590.
Zaitsev, S. I. (1992), *Physica* 189A: 441.
Zhang, Y.-C. (1989), *Phys. Rev. Lett.* 63: 470.
Zieve, R. J., Rosenbaum, T. F., Jaeger, H. M., Seidler, G. T., Crabtree, G. W., and Welp, U. (1996), *Phys. Rev. B* 53: 11849.

Index

Printed in the United States
By Bookmasters